THE PRINCIPLES OF

ELECTROMAGNETIC THEORY

THE PRINCIPLES OF
ELECTROMAGNETIC THEORY

ATTAY KOVETZ

Professor of Physics at Tel Aviv University

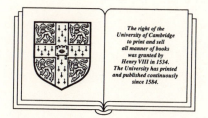

CAMBRIDGE UNIVERSITY PRESS

Cambridge

New York   Port Chester

Melbourne   Sydney

Published by the Press Syndicate of the University of Cambridge
The Pitt Building, Trumpington Street, Cambridge CB2 1RP
40 West 20th Street, New York NY 10011, USA
10 Stamford Road, Oakleigh, Melbourne 3166, Australia

© Cambridge University Press 1990

First published 1990

Printed in Great Britain at the University Press, Cambridge

British Library cataloguing in publication data

Kovetz, Attay
    The principles of electromagnetic theory.
    1. Electromagnetism. Theories
    I. Title
    5330.14

Library of Congress cataloguing in publication data available

ISBN 0 521 39106 7 hard covers
ISBN 0 521 39997 1 paperback

UPM

To the memory of
my father

Contents

Preface

This book is the outcome of a course entitled 'Analytical Electromagnetism', which I have taught for a number of years at Tel Aviv University. The course is attended by physics students during the second semester of their second year undergraduate studies. It is on an intermediate level: the students have all been exposed to electromagnetism, as well as to special relativity, through several courses (both in the class and in the laboratory) on classical and modern physics; and all graduate students have to take a further course called 'Advanced Electromagnetism'.

Obviously, then, the students who attend such an intermediate course have all seen Maxwell's equations in one form or another. They have all heard of Poynting's vector, electric displacement, Joule heating, Lorenz force and electromagnetic waves. But, despite this familiarity, the average student is uncertain about the concepts and laws of electromagnetic theory. Do Maxwell's equations hold in all frames? If not, how are the laws of electromagnetism to be stated in those frames in which the equations do *not* hold? Is $\mathbf{j} \cdot \mathbf{E}$ a density of mechanical power, of heating, or perhaps neither? Is the force on a piece of dielectric material the (exact? approximate?) resultant of the Lorenz forces acting on the individual charges (or dipoles)? If not, why? If it is, where do the derivatives of the permittivity *at constant temperature* come from? If the expression for electric energy is obtained by calculating the mechanical work done by the *conservative*, electrostatic field, and the magnetic energy by a similar calculation that assumes $\mathbf{curl}\ \mathbf{E} = -\partial\mathbf{B}/\partial t \neq 0$, does it make any sense to add the two and regard the sum as the total electromagnetic energy?

Is the result, however arrived at, an energy in the mechanical sense (subject to changes determined by work) or in the thermodynamical sense (subject to changes determined by work or heat)? In the latter case, is it really total, or perhaps only *free*, energy? Should we expect Maxwell's stress tensor to be symmetric? Is it? Does it matter? Is $D = \epsilon_0 E + P$ a constitutive relation, a definition, a universal law? Does it hold in moving bodies? Such questions, and many others, baffle most students of electromagnetism. Unless there are some gaping faults in the logical structure of electromagnetic theory, the reason for these difficulties must surely lie in the way in which this theory is usually presented.

The difficulties with electromagnetism in arbitrary frames are, in fact, the result of an incomplete statement of the fundamental principles. In courses on classical mechanics the student is told that the basic laws hold in inertial frames, connected to each other by Galilei transformations; that forces are absolute (invariant), but accelerations are not; and that this difference in behaviour with respect to change of frame leads to the appearance of fictitious forces in frames that are not inertial. By contrast, the basic laws of electromagnetism are usually only stated in Lorenz frames (the inertial frames of relativity), as though the student, who has successfully worked through numerous exercises involving fictitious forces in mechanics, is not to be trusted with information regarding the *general* transformation properties of electromagnetic quantities. He can, of course, find the answer in one of the late chapters of a book on general relativity, but then he is apt to give up the search before realizing that the answer has really nothing to do with gravitation.

The difficulties that the student of electromagnetism encounters in connection with the concepts of force and energy are of a more serious kind. Force and energy are subject to the laws of mechanics and thermodynamics, and their discussion requires the establishment of a definite relationship between these disciplines and electromagnetism. About this relationship the textbooks are only too often strangely vague. The treatment is usually wordy, in the worst thermodynamic tradition; in fact, the laws of mechanics and thermodynamics are seldom stated in mathematical form. And the manner in which these laws must be modified in order to account for both electromagnetic momentum *and* energy flux remains unclear. It is hardly surprising that the student is left wondering whether ponderomotive forces can ever be calculated for situations that are far removed from equilibrium, or whether radiation — especially 'heat radiation' — is, or is not, a form of heat.

I believe that an intermediate course affords a good opportunity for presenting electromagnetism as a coherent and logical theory. This is the purpose of this book. Its level is roughly that of other texts on electromagnetism that start from Maxwell's equations. In particular, it assumes that the reader requires little or no motivation for the introduction of various electromagnetic concepts and definitions. But it differs from most other texts in that it presents electromagnetism as a *classical* theory, based on principles that are independent of the atomic constitution of matter. Whereas the principles of classical mechanics are never claimed to depend on the fact that ordinary (terrestrial) matter consists of atoms, textbooks on electromagnetism abound with discussions of stretched molecules forming tiny dipoles and of Amperian currents associated with atomic electrons, all dating back to Lorenz's eighty years old Theory of Electrons. Not only is the student asked to believe that the average value of a quantity can be found in the absence of any knowledge regarding its distribution. He is also expected to forget that the existence of atoms can only be established on the basis of quantum mechanics, using the very Hamiltonian operators that are suggested by electromagnetism. Obviously, then, the contemporary author, who no longer believes the classical theory of electrons — and who looks to the solid state physicists for an explanation of its well-known successes — is not being quite frank with his readers. Clearly, a presentation of electromagnetic theory that avoids the atomic constitution of matter is to be preferred on both didactic and logical grounds.

The outline of the book is as follows: the laws of electromagnetism are stated, in the first three chapters, in tensor form. Besides taking care of questions regarding transformations and invariance, the tensorial formulation has the added advantage that the students will not have to re-learn electromagnetism when they take a course in general relativity. For, although curved space-time in not mentioned in these chapters, the possibility is nowhere denied.

Chapters IV and V establish the link between electromagnetism, mechanics and thermodynamics. The principles of thermodynamics are stated in mathematical terms (the second in the form of the Clausius-Duhem inequality), and the complete set of electromagnetic, mechanic and thermodynamic laws is then treated in a manner which is a direct generalization of the method that Coleman and Noll introduced (in 1963) in order to establish the link between mechanics and thermodynamics on a rational basis. Maxwell stress tensors, ponderomotive forces, equations of motion and of energy and many other results all follow by a process of mathematical, and therefore logical, deduction. Charged particles are discussed separately and treated classically, as well as relativistically.

The remaining six chapters deal with applications of the theory to electrostatics, dielectrics, magnetic materials, conductors and radiation. To some degree, the choice of subjects invariably reflects a personal preference. I have tried to convince the reader that electromagnetism is a theory, not only of phenomena, but also of materials, and that it can be readily applied to moving, as well as to stationary, bodies. I have also attempted to dispel some common prejudices (such as the belief, which most students share, that electromagnetism is a linear theory) and misconceptions (for example, that the theory needs to be supplemented by a principle of 'causality'). I have left out the treatment of alternating-current circuits because they are adequately treated in other courses (but there is a section on the skin effect).

Many of the problems scattered throughout the book are really parts of a mathematical proof or derivation that have been left as exercises for the reader. The remaining problems, representing applications of the theory, are of the kind usually found at the end of a textbook chapter.

The prerequisites for a student who intends to use this book as a text for an intermediate or advanced course on electromagnetic theory should present no problem. The mathematical tools required are no more than vector calculus and Legendre polynomials. Tensors are introduced for the purpose of formulating the basic laws of electromagnetism in the most general manner, but the student who is not especially interested in questions of invariance need only take notice of the definitions and statements. A working knowledge of tensors is not required; beyond the first three chapters tensors are not even mentioned. Of the physics curriculum, besides a basic, introductory course on electromagnetism, some familiarity with the basic concepts of analytical mechanics, special relativity and thermodynamics is assumed, but any of these can be studied concurrently with a reading of this book.

I would like to thank my wife Dina (née Prialnik) for her constant encouragement and for a critical reading of the manuscript.

Tel Aviv 1989 A. K.

Notation

In some cases I have found it necessary to depart from the notation that has become standard in treatments of electromagnetism. Of these departures, the most important are:

Electric charge $\quad Q, e$

Charge density $\quad q$

Mass density $\quad \rho$

Stress tensor $\quad T$

Temperature $\quad \vartheta$.

In each section, equations – or groups of equations – are numbered consecutively. Thus, for example, (10.6) denotes the sixth equation of §10; and if this is a group of equations, $(10.6)_2$ denotes its second member.

CHAPTER I

Electric Charges

and Currents

1. Charge conservation

Electric charge is a property of bodies in nature. Logically speaking, it is a primitive: its status in electromagnetism is like that of point or plane in geometry, mass or force in mechanics, energy or temperature in thermodynamics, etc. For the time being we shall be concerned with those cases in which the charge Q is smoothly distributed over volume; more precisely, those situations in which it is possible to define a charge density q such that

$$Q = \int q \, dV, \tag{1.1}$$

where Q is the charge residing in some region and the integral is over the volume of that region. Of course, Q and q may each be positive, negative or zero.

The fundamental assumption we shall make regarding electric charge is that it is conserved. By that we mean that if the charge within a certain region has changed, it is because charge has passed out of the region, or entered into it, through the bounding surface:

$$\Delta Q = \Delta \int q \, dV = - \int_{t_1}^{t_2} dt \oint \alpha \, dS. \tag{1.2}$$

Here $\Delta Q = Q(t_2) - Q(t_1)$ is the change in charge over the time interval from t_1 to t_2. The last integral, which extends over the complete surface, is the net rate at which electric charge leaves the region: $\alpha \, dS$ is

1

the rate at which charge flows out through the surface element dS. The function α, the existence of which is part of the principle (or axiom) of charge conservation, depends on position $\mathbf{x}$ and time t, and on properties of the surface. But a theorem of Cauchy (generalized by Hamel and Noll) in continuum mechanics imposes severe restrictions on the latter dependence: α has a common value for all surface elements dS having a common tangent plane. It is therefore a function only of the orientation of the tangent plane to the surface (and not of the curvature, for example). Since the orientation of the tangent plane is determined by the components of the unit normal $\mathbf{n}$, α is a function of $\mathbf{x}$, t and $\mathbf{n}$. Finally, by Cauchy's theorem, the dependence of α on the components n_i of $\mathbf{n}$ is linear and homogeneous:

$$\alpha(\mathbf{x}, t, \mathbf{n}) = j_1(\mathbf{x}, t)n_1 + j_2(\mathbf{x}, t)n_2 + j_3(\mathbf{x}, t)n_3. \qquad (1.3)$$

The coefficients j_i are the components of a three-dimensional vector†
$\mathbf{j}$. Equation (1.3) can then be written in the form

$$\alpha = \mathbf{j} \cdot \mathbf{n} = j_n(\mathbf{x}, t). \qquad (1.4)$$

According to (1.2), the amount of charge passing (in the direction of $\mathbf{n}$), per unit time, through the surface element dS is $\mathbf{j} \cdot \mathbf{n}\, dS$. The vector $\mathbf{j}$ is called the *current density*.

Obviously, there are two ways of defining the unit normal on a surface. We choose the direction pointing out of the volume. Then

$$\oint \alpha\, dS = \oint \mathbf{j} \cdot \mathbf{n}\, dS = \int \operatorname{div} \mathbf{j}\, dV, \qquad (1.5)$$

and the principle of charge conservation becomes

$$\frac{d}{dt} \int q\, dV + \int \operatorname{div} \mathbf{j}\, dV = 0. \qquad (1.6)$$

In local form:

$$\frac{\partial q}{\partial t} + \operatorname{div} \mathbf{j} = 0. \qquad (1.7)$$

---

† This is an example of the use of a test for vector character (cf. Problem 2–6): $\mathbf{n}$ is a vector, and $\alpha = \mathbf{j} \cdot \mathbf{n}$ a scalar; hence $\mathbf{j}$ is a vector.

If we denote the dimensions of length, time and charge by L, T and C, respectively, the dimensions of charge density q are C/L^3, and the dimensions of current density $\mathbf{j}$ are $C/(TL^2)$.

The notation div $\mathbf{j}$ stands for $\partial j_i/\partial x^i$, where summation over the repeated index i is understood. We shall now write the principle of charge conservation in four-dimensional notation. First, we add the time as a fourth coordinate, x^4, to the three space coordinates x^i, $i = 1, 2, 3$:

$$(x^\alpha) = (x^1, x^2, x^3, x^4) = (x^i, x^4) = (\mathbf{x}, t); \tag{1.8}$$

we use Latin indices (like i) for the three space coordinates and Greek indices (like α) for the four space-time coordinates. Accordingly, a repeated Latin index implies summation from 1 to 3, and a Greek index, from 1 to 4. A point $(x^\alpha) = (\mathbf{x}, t)$ in space-time is called an *event*.

We now define

$$(\sigma^\alpha) = (\sigma^i, \sigma^4) = (\mathbf{j}, q) \tag{1.9}$$

and call σ^α the *charge-current density*. With this notation, $\partial q/\partial t = \partial \sigma^4/\partial x^4$ and div $\mathbf{j} = \partial \sigma^i/\partial x^i$, so that (1.7) becomes

$$\frac{\partial \sigma^\alpha}{\partial x^\alpha} = 0. \tag{1.10}$$

As we have already noted, this is the local form of (1.2), which itself can now be written in the form

$$\int \int \frac{\partial \sigma^\alpha}{\partial x^\alpha} \, dV \, dt = \int \frac{\partial \sigma^\alpha}{\partial x^\alpha} \, d^4x = 0. \tag{1.11}$$

Our next object is to arrive at a formulation of the principle of charge conservation that is independent of the coordinates we happen to be using for labelling space-time. After all, charge conservation as we have originally formulated it has a meaning which obviously transcends any choice of space-time coordinates.

2. Transformations and tensors

Let

$$x^{\alpha'} = x^{\alpha'}(x^\alpha) \tag{2.1}$$

be a coordinate transformation. We shall assume that the functions that appear in (2.1) possess a sufficient number of derivatives, and that the Jacobian (x'/x) of the transformation does not vanish. This ensures that the transformation is invertible. A set of functions of the coordinates that obeys the transformation law

$$f^{\alpha'\beta'\cdots}_{\gamma'\cdots}(x') = \frac{\partial x^{\alpha'}}{\partial x^{\alpha}} \frac{\partial x^{\beta'}}{\partial x^{\beta}} \cdots \frac{\partial x^{\gamma}}{\partial x^{\gamma'}} \cdots f^{\alpha\beta\cdots}_{\gamma\cdots}(x) \qquad (2.2)$$

is called a *tensor*. The number of indices (upper *and* lower) is called the *rank* or *order* of the tensor. Upper indices are called *contravariant*; there is one partial derivative $\partial x^{\alpha'}/\partial x^{\alpha}$ on the right hand side of (2.2) – with a summation over α – for each contravariant index α' on the left hand side. Lower indices are called *covariant*; there is one partial derivative $\partial x^{\gamma}/\partial x^{\gamma'}$ on the right hand side of (2.2) – with a summation over γ – for each covariant index γ' on the left hand side.

The simplest tensor is a *scalar* ϕ, with no indices at all. The set of functions then consists of a single member, or *component*, and the law of transformation reduces to

$$\phi(x') = \phi(x). \qquad (2.3)$$

As an example of a scalar, we cite the temperature. In classical thermodynamics the temperature at an event $(x) = (\mathbf{x}, t)$ is assumed to be independent of the labels that are used to denote that event. It is thus assumed to satisfy (2.3) and is therefore a scalar.

According to (2.1), the differentials $dx^{\alpha'}$ are given, in terms of the differentials of the unprimed coordinates, by

$$dx^{\alpha'} = \frac{\partial x^{\alpha'}}{\partial x^{\alpha}} dx^{\alpha}. \qquad (2.4)$$

Comparison with (2.2) shows that the four differentials of the coordinates are the components of a contravariant tensor of rank one.

If we differentiate (2.3) with respect to $x^{\alpha'}$ and use the chain rule, we obtain

$$\frac{\partial \phi}{\partial x^{\alpha'}} = \frac{\partial x^{\alpha}}{\partial x^{\alpha'}} \frac{\partial \phi}{\partial x^{\alpha}}. \qquad (2.5)$$

Thus the four components of the (four-dimensional) gradient $\partial \phi/\partial x^{\alpha}$ of a scalar ϕ constitute a covariant tensor of rank one.

A tensor of rank one is called a *vector*. Vectors can be contravariant (like dx^α) or covariant (like $\partial\phi/\partial x^\alpha$).

In classical mechanics we often deal with the restricted class of three-dimensional transformations that represent rotations and translations of Cartesian frames. Such transformations are of the form

$$x^{i'} = A^{i'}_i (x^i - a^i),\qquad\qquad(2.6)$$

where $A^{i'}_i$ is an orthogonal matrix. The elements of $A^{i'}_i$ (not a tensor) and the three a^i (not a vector) are independent of $\mathbf{x}$, so that (2.6) is linear. In matrix notation, we have $\mathbf{x}' = A(\mathbf{x} - \mathbf{a})$ and $\mathbf{x} = A^T\mathbf{x}' + \mathbf{a}$, where A^T denotes the transpose of A. It follows that both $\partial x^{i'}/\partial x^i$ and $\partial x^i/\partial x^{i'}$ are equal to $A^{i'}_i$. Thus, with respect to the class of Cartesian transformations (2.6), contravariant and covariant tensors transform in the same way. That is why we never bother to distinguish between contravariance and covariance in connection with Cartesian tensors (such as Euler's tensor of inertia for a rigid body). For vectors, in particular, we can use the notation of ordinary vector calculus and write $\mathbf{a}$ for a_i or a^i. But the distinction does arise, even in classical mechanics, as soon as we contemplate transformations between *moving* frames, such as the *Galilei transformations*

$$x^{r'} = A^{r'}_r (x^r - u^r x^4) + \text{const.,}$$

$$x^{4'} = x^4 + \text{const.,}\qquad\qquad(2.7)$$

where $A^{r'}_r$ is an orthogonal matrix with constant elements, and the constants u^r are the components of the velocity of the frame (x') with respect to the frame (x). The transformation (2.7) is four-dimensional (a special case of (2.1)). Covariant and contravariant four-dimensional tensors will transform differently under (2.7), because now $\partial x^{r'}/\partial x^4 = -A^{r'}_r u^r$, whereas $\partial x^4/\partial x^{r'} = 0$.

PROBLEMS

The following problems contain some easy theorems on tensors.

Problem 2–1. If $f_{\alpha'} = (\partial x^\alpha/\partial x^{\alpha'})f_\alpha$ and $f_{\alpha''} = (\partial x^{\alpha'}/\partial x^{\alpha''})f_{\alpha'}$, then $f_{\alpha''} = (\partial x^\alpha/\partial x^{\alpha''})f_\alpha$. This is the transitive property of the tensor transformation law.

Problem 2–2. The sum of tensors of the same type is a tensor.

Problem 2–3. If f and g are tensors of the same type which are equal in one coordinate system, they are equal in all other coordinate systems.

Problem 2–4. If $A_{\alpha\beta}$ is a tensor then $A_{(\alpha\beta)} = A_{\alpha\beta} + A_{\beta\alpha}$ and $A_{[\alpha\beta]} = A_{\alpha\beta} - A_{\beta\alpha}$ are tensors. The tensor transformation preserves symmetry.

Problem 2–5. If $f_{\beta\gamma\delta\ldots}^{\alpha\cdots}$ is a tensor of rank r, then $f_{\alpha\gamma\delta\ldots}^{\alpha\cdots}$ (summation over α) is a tensor of rank $r - 2$.

Problem 2–6. If $A_\alpha X^\alpha$ is a scalar for arbitrary contravariant vectors X^α, then A_α is a covariant vector.

Problem 2–7. If u^α and v^α are contravariant vectors, then $f^{\alpha\beta} = u^\alpha v^\beta$ is a contravariant tensor of rank two.

The tensors defined by (2.2) are actually called *absolute* tensors. A somewhat more general class of tensors is obtained if (2.2) is replaced by

$$f_{\gamma'\ldots}^{\alpha'\beta'\cdots}(x') = |(x'/x)|^{-w} \frac{\partial x^{\alpha'}}{\partial x^\alpha} \frac{\partial x^{\beta'}}{\partial x^\beta} \cdots \frac{\partial x^\gamma}{\partial x^{\gamma'}} \cdots f_{\gamma\ldots}^{\alpha\beta\cdots}(x). \qquad (2.8)$$

This transformation law differs from (2.2) in that the absolute value of the Jacobian, raised to the power $-w$, appears as a common factor on the right hand side. A tensor that obeys (2.8) is called a *relative tensor of weight w*. Relative tensors share many properties with absolute tensors (which are, of course, relative tensors of weight zero).

PROBLEM

Problem 2–8. Using the transitivity property $(x/x'') = (x/x')(x'/x'')$ of Jacobian determinants, show that each one of the theorems contained in Problems 2–1 to 2–7 can be modified so as to apply to relative tensors.

In what follows we shall only need to consider contravariant relative tensors of weight 1. Such tensors are called *tensor densities*. The distinction between absolute tensors and tensor densities becomes necessary when we have to deal with integrals, as in (1.11). This is because

$$\int d^4x = \int |(x'/x)|^{-1} d^4x'. \qquad (2.9)$$

As a final generalization, we consider the transformation law

$$f_{\gamma'\cdots}^{\alpha'\cdots}(x') = |(x'/x)|^{-w}\mathrm{sgn}\,(x'/x)\frac{\partial x^{\alpha'}}{\partial x^\alpha}\cdots\frac{\partial x^\gamma}{\partial x^{\gamma'}}\cdots f_{\gamma\cdots}^{\alpha\cdots}(x), \qquad (2.10)$$

which differs from (2.8) by the sign of the Jacobian. Tensors that obey (2.10) are called *axial*, or *oriented*, tensors.

Of course, the differences between absolute, relative and axial tensors disappear when we consider those coordinate transformations for which $(x'/x) = 1$. Similarly, for transformations with positive Jacobian, there is no difference between absolute and axial tensors, or between tensor densities and axial tensor densities.

In electromagnetism we only have to deal with tensors of rank two at most. For reasons which will become clear later on, it will also turn out that the important coordinate transformations are all linear, with $(x'/x) = 1$. For tensors of low rank, and for such simple coordinate transformations, the tensor transformation law becomes far less formidable than the general relation (2.10) might suggest. Although tensors are indispensable for the formulation of the principles of electromagnetism, we shall find that these very principles will enable us to cast the theory entirely in terms of three-dimensional vectors.

3. Transformation of the charge-current density

We shall now assume that σ is a vector density. Its transformation law is then

$$\sigma^{\alpha'}(x') = |(x'/x)|^{-1}\frac{\partial x^{\alpha'}}{\partial x^\alpha}\sigma^\alpha(x). \qquad (3.1)$$

Accordingly, we have

$$\frac{\partial\sigma^{\alpha'}}{\partial x^{\alpha'}} = \frac{\partial}{\partial x^{\alpha'}}\left[|(x'/x)|^{-1}\frac{\partial x^{\alpha'}}{\partial x^\alpha}\sigma^\alpha\right]$$

$$= |(x'/x)|^{-1}\frac{\partial\sigma^\alpha}{\partial x^\alpha} + \sigma^\alpha\frac{\partial}{\partial x^{\alpha'}}\left[|(x'/x)|^{-1}\frac{\partial x^{\alpha'}}{\partial x^\alpha}\right]. \qquad (3.2)$$

PROBLEMS

Problem 3–1. $A(t)$ is a matrix with elements that are functions of a variable t.

$\dot{A}(t)$ denotes the matrix with elements that are the derivatives, with respect to t, of the elements of A. Prove the formula for the derivative of a determinant:

$$\frac{d}{dt}(\det A) = (\det A)\operatorname{tr} A^{-1}\dot{A}, \tag{3.3}$$

where $\operatorname{tr} A^{-1}\dot{A}$ is the trace of the matrix product $A^{-1}\dot{A}$.

Problem 3–2. Prove that

$$\frac{\partial(x'/x)}{\partial x^\alpha} = (x'/x)\frac{\partial x^\beta}{\partial x^{\alpha'}}\frac{\partial^2 x^{\alpha'}}{\partial x^\alpha \partial x^\beta}. \tag{3.4}$$

Problem 3–3. Prove that

$$\frac{\partial}{\partial x^{\alpha'}}\left[|(x'/x)|^{-1}\frac{\partial x^{\alpha'}}{\partial x^\alpha}\right] = 0. \tag{3.5}$$

According to (3.5), the last term in (3.2) vanishes. Thus

$$\frac{\partial\sigma^{\alpha'}}{\partial x^{\alpha'}} = |(x'/x)|^{-1}\frac{\partial\sigma^\alpha}{\partial x^\alpha}, \tag{3.6}$$

and $\partial\sigma^\alpha/\partial x^\alpha = 0$ implies $\partial\sigma^{\alpha'}/\partial x^{\alpha'} = 0$. Our assumption, that σ is a relative vector, therefore ensures the invariance of the principle of charge conservation in its local form (1.10), as well as in its global, or integral, form

$$\int\frac{\partial\sigma^{\alpha'}}{\partial x^{\alpha'}}\,d^4x' = \int\frac{\partial\sigma^\alpha}{\partial x^\alpha}|(x'/x)|^{-1}\,d^4x' = \int\frac{\partial\sigma^\alpha}{\partial x^\alpha}\,d^4x = 0 \tag{3.7}$$

(cf. (2.9)). The assumption (3.1) is, in fact, equivalent to the invariance of the principle of charge conservation.

The laws of classical mechanics are known to be invariant with respect to the Galilei transformation (cf. (2.7))

$$x^{r'} = A_r^{r'}(x^r - u^r x^4) + \text{const.},$$

$$x^{4'} = x^4 + \text{const.};$$

for these transformations $(x'/x) = \det A = \pm1$, so that $|(x'/x)| = 1$. Using (3.1), we can now find the transformation formulae for the components of $\sigma = (\mathbf{j}, q)$ under the Galilei transformation:

$$j^{r'} = \sigma^{r'} = A_r^{r'}(\sigma^r - u^r\sigma^4) = A_r^{r'}(j^r - qu^r),$$

$$q' = \sigma^{4'} = \sigma^4 = q. \tag{3.8}$$

It is clear that $q' = q$ and that the components $j^{r'}$ are the components $j^r - qu^r$, rotated by the matrix A. In three-dimensional vector notation we can write

$$\mathbf{j'} = \mathbf{j} - q\mathbf{u},$$

$$q' = q. \tag{3.9}$$

These are the transformation laws of the charge and current densities under a Galilei transformation.

It should, perhaps, be noted that our four-dimensional notation has nothing to do with the theory of relativity. In fact, the only use we have made of it was in connection with the Galilei transformation. Nor does the principle of electric charge conservation have anything to do with relativity.

4. The charge-current potential

Let $f^{\alpha\beta}$ be an antisymmetric tensor density. The vector

$$\sigma^\alpha = \frac{\partial f^{\alpha\beta}}{\partial x^\beta}, \qquad \alpha = 1, 2, 3, 4, \tag{4.1}$$

will satisfy the charge conservation law (1.10). The converse is also true: if σ satisfies (1.10), there exists an antisymmetric f such that (4.1) holds. This is a generalization, to four dimensions, of the statement, 'if div $\mathbf{b} = 0$ then there exists an $\mathbf{a}$ such that $\mathbf{b} = \mathbf{curl}\ \mathbf{a}$'. But f is not uniquely determined by (4.1): we can add to it the sum $\partial g^{\alpha\beta\gamma}/\partial x^\gamma$, where $g^{\alpha\beta\gamma}$ is antisymmetric in its three indices, but is otherwise arbitrary.

PROBLEM

Problem 4–1. Prove that (4.1) is a tensor equation.

Since the charge-current density σ is obtained from f by differentiation, f is called a *charge-current potential*. Of its sixteen components, the four along the diagonal vanish; of the remaining twelve components, the six above the diagonal are opposite in sign to the six below. Thus there are only six independent components. We shall divide them into two groups of three:

$$D^r = f^{4r}, \qquad \epsilon^{rst} H_t = f^{rs}; \tag{4.2}$$

here ϵ is the permutation symbol ($\epsilon^{123} = 1$). We note the following two examples of the use of ϵ: if $\mathbf{c} = \mathbf{a} \times \mathbf{b}$ then $c^r = \epsilon^{rst} a_s b_t$; if $\mathbf{a} = \mathbf{curl}\ \mathbf{b}$ then $a^r = \epsilon^{rst}(\partial/\partial x^s) b_t$. With the notations (4.2), f becomes

$$f = \begin{pmatrix} 0 & H_3 & -H_2 & -D^1 \\ -H_3 & 0 & H_1 & -D^2 \\ H_2 & -H_1 & 0 & -D^3 \\ D^1 & D^2 & D^3 & 0 \end{pmatrix}. \tag{4.3}$$

Of course, the mere fact that we denote the components of f by $D^1, \ldots, H_3$ does not mean that D^r is a contravariant vector, or that H_r is a covariant vector. A subset of the components of a tensor is not itself a tensor, but it may transform as a tensor (of lower order) with respect to transformations in a subspace. This is indeed what happens in the present case.

PROBLEMS

Problem 4–2. Use the definition of a determinant to show that

$$\epsilon^{rst} A_r^u A_s^v A_t^w = \epsilon^{uvw} \det A. \tag{4.4}$$

Problem 4–3. By substituting the Jacobian matrix for A in (4.4), prove that ϵ is an axial tensor density.

Problem 4–4. From the transformation law of the tensor density f, show that, with respect to the subset of space transformations $x^{r'} = x^{r'}(x^r), x^{4'} = x^4, D^r$ is a three-dimensional vector density and H_r is an axial covariant three-dimensional vector.

Let us now subject the tensor density f to the Galilei transformation (2.7). If we use the notation (4.2) for the components of f, the result (which is analogous to (3.9)) is

$$\mathbf{D'} = \mathbf{D},$$

$$\mathbf{H'} = \pm(\mathbf{H} - \mathbf{u} \times \mathbf{D}), \tag{4.5}$$

where the $\pm$ in the last equation refers to the sign of det A. Except when otherwise mentioned, we shall confine ourselves to *proper* transformations, for which det $A = 1$, and use the $+$ sign in (4.5). We see that, under a Galilei transformation, $\mathbf{D}$ is unchanged (except for the rotation induced by the orthogonal matrix A), but $\mathbf{H}$ changes.

We shall find it convenient to use the abbreviated notation

$$f = \{\mathbf{H}, \mathbf{D}\} \tag{4.6}$$

instead of (4.3).

We now write out (4.1), using the new names we have introduced for the components of f:

$$\sigma^4 = \frac{\partial f^{4\beta}}{\partial x^\beta} = \frac{\partial f^{4r}}{\partial x^r} = \frac{\partial D^r}{\partial x^r},$$

$$\sigma^r = \frac{\partial f^{r\beta}}{\partial x^\beta} = \frac{\partial f^{rs}}{\partial x^s} + \frac{\partial f^{r4}}{\partial x^4} = \epsilon^{rst}\frac{\partial H_t}{\partial x^s} - \frac{\partial D^r}{\partial x^4}. \tag{4.7}$$

In three-dimensional notation, remembering $\sigma = (\mathbf{j}, q)$, we have

$$\operatorname{div}\mathbf{D} = q,$$

$$\operatorname{curl}\mathbf{H} - \frac{\partial \mathbf{D}}{\partial t} = \mathbf{j}. \tag{4.8}$$

These are two of Maxwell's equations, and they express the principle of charge conservation in terms of the charge-current potential $f = \{\mathbf{H}, \mathbf{D}\}$. Since these equations are merely the tensor equation (4.1), expressed in a different notation, they hold in all frames. We shall always use Maxwell's equations (4.8), rather than (4.1), in applications of electromagnetism. In the next chapter we shall introduce two further principles. These, too, will first be stated in terms of tensors and then expressed in three-dimensional language.

The usual method of showing that $f = \{\mathbf{H}, \mathbf{D}\}$ is not uniquely determined by $\sigma = (\mathbf{j}, q)$ is to exhibit non-trivial solutions of the homogeneous system

$$\text{div}\,\mathbf{D} = 0,$$

$$\mathbf{curl}\,\mathbf{H} - \frac{\partial \mathbf{D}}{\partial t} = 0. \tag{4.9}$$

The first of these is satisfied by $\mathbf{D} = \mathbf{curl}\,\mathbf{a}$ with arbitrary $\mathbf{a}$; the second by $\mathbf{H} = \partial\mathbf{a}/\partial t + \mathbf{grad}\,\psi$ with arbitrary ψ. We shall need a special principle to fix f, that is, $\mathbf{H}$ and $\mathbf{D}$.

The non-uniqueness presents no problem when we use (4.1) or (4.8) to calculate the charge-current density $\sigma = (\mathbf{j}, q)$ from the potential $f = \{\mathbf{H}, \mathbf{D}\}$. This is, in fact, the way in which these equations are used in material media. The response functions of materials specify the charge-current potential, and the foregoing equations are then used to calculate the so-called polarization and magnetization charge and current densities.

It is clear that the dimensions of $\mathbf{D}$ and $\mathbf{H}$ are C/L^2 and $C/(LT)$, respectively.

PROBLEM

Problem 4–5. The *dual* of f is defined by

$$(\text{dual}\,f)_{\alpha\beta} = \frac{1}{2}\epsilon_{\alpha\beta\gamma\delta}f^{\gamma\delta},$$

where $\epsilon_{\alpha\beta\gamma\delta}$ is the four-dimensional permutation symbol. Show that

$$\frac{1}{4}f^{\alpha\beta}(\text{dual}\,f)_{\alpha\beta} = -D^r H_r.$$

5. Integral laws and jump conditions

If we integrate $(4.8)_1$ over some volume and use Gauss's formula† to transform to an integral over the boundary surface, we obtain

$$\oint D_n\,dS = \int q\,dV = Q, \tag{5.1}$$

† The correct attribution is to Green.

where D_n is the component of $\mathbf{D}$ in the direction of $\mathbf{n}$. The charge in any volume is therefore equal to the flux of $\mathbf{D}$ through the boundary surface.

We now calculate the flux of either side of $(4.8)_2$ through a fixed, open, surface. Use of Stokes's formula† gives

$$\oint_c H_s \, ds = \int j_n \, dS + \frac{d}{dt} \int D_n \, dS. \tag{5.2}$$

The line integral on the left is taken along the curve c that constitutes the boundary of the open surface, in the sense that would cause a right-handed screw to advance in the direction of the $\mathbf{n}$ that appears on the right hand side of (5.2); and H_s is the component of $\mathbf{H}$ along the tangent to the curve (in the direction corresponding to the foregoing sense along c). This line integral is called the *circulation* of $\mathbf{H}$ around c. In the second term on the right we have used the fact that the surface is fixed in order to place the time differentiation in front of the integral.

There is a more general formula than (5.2) which applies to a moving (and deforming) surface. Let $\mathbf{v}(\mathbf{x}, t)$ be a differentiable velocity field (not necessarily the velocity of any actual material). Clearly,

$$\mathbf{curl}\,(\mathbf{H} - \mathbf{v} \times \mathbf{D}) = \mathbf{j} + \frac{\partial \mathbf{D}}{\partial t} - \mathbf{curl}\,(\mathbf{v} \times \mathbf{D})$$

$$= \mathbf{j} - \mathbf{v}\,\mathrm{div}\,\mathbf{D} + \frac{\partial \mathbf{D}}{\partial t} + \mathbf{v}\,\mathrm{div}\,\mathbf{D} - \mathbf{curl}\,(\mathbf{v} \times \mathbf{D})$$

$$= \mathbf{j} - q\mathbf{v} + \overset{*}{\mathbf{D}}, \tag{5.3}$$

where, for any vector $\mathbf{A}$ and velocity field $\mathbf{v}$,

$$\overset{*}{\mathbf{A}} = \frac{\partial \mathbf{A}}{\partial t} + \mathbf{v}\,\mathrm{div}\,\mathbf{A} - \mathbf{curl}\,(\mathbf{v} \times \mathbf{A}). \tag{5.4}$$

PROBLEM

Problem 5–1. Let $\int A_n \, dS$ be the instantaneous flux of $\mathbf{A}$ through an open

† The correct attribution is to Kelvin.

surface which is swept along by a velocity field $\mathbf{v}$. Prove that

$$\frac{d}{dt} \int A_n \, dS = \int \overset{*}{\mathbf{A}} \cdot \mathbf{n} \, dS. \tag{5.5}$$

From (5.3)–(5.5) we obtain

$$\oint_c (\mathbf{H} - \mathbf{v} \times \mathbf{D})_s \, ds = \int (\mathbf{j} - q\mathbf{v})_n \, dS + \frac{d}{dt} \int D_n \, dS.$$

We have obtained the integral laws

$$\oint D_n \, dS = \int q \, dV,$$

$$\oint_c (\mathbf{H} - \mathbf{v} \times \mathbf{D})_s \, ds = \int (\mathbf{j} - q\mathbf{v})_n \, dS + \frac{d}{dt} \int D_n \, dS, \tag{5.6}$$

from Maxwell's equations (4.8). Conversely, these integral laws lead to Maxwell's equations wherever the integrands are differentiable. But the integral laws may be meaningful even when div $\mathbf{D}$ or **curl H** do not exist, or when electric charges are distributed over surfaces, along lines or concentrated in points, or when electric currents are flowing along surfaces or lines. Such cases are important in electromagnetism, and a good theory should not exclude them.

From now on we shall assume that all relevant quantities are piecewise smooth. By this we mean that they are continuously differentiable up to any desired order everywhere, except on elements of discontinuity, which may be points, lines or surfaces; and that the number of such elements in any closed subdomain of space is finite. For piecewise smooth quantities the integral laws (5.6) are everywhere meaningful, and from now on we shall take them, rather than the differential equations, to be the basic laws.†

† We refrain at this stage from identifying $(5.6)_1$ with Gauss's law, or $(5.6)_2$ with Ampère's law (for moving and deforming surfaces). That is because $\mathbf{D}$ and $\mathbf{H}$ – the components of the charge-current potential f – are as yet unrelated to the electric and magnetic fields $\mathbf{E}$ and $\mathbf{B}$. These fields and this relation are the subjects of two further principles of electromagnetism, which we shall introduce in the next chapter.

Note that every integrand in (5.6) is Galilei-invariant. We know that this is so for q and $\mathbf{D}$ ((3.9)$_2$ and (4.5)$_1$). Consider next the two quantities

$$\mathcal{J} = \mathbf{j} - q\mathbf{v},$$

$$\mathcal{H} = \mathbf{H} - \mathbf{v} \times \mathbf{D}. \tag{5.7}$$

We call $\mathcal{J}$ the *conduction current density*, because it is the part of $\mathbf{j}$ that is not due to the *convection* of charge density q at velocity $\mathbf{v}$. In order to check the Galilei-invariance of $\mathcal{J}$, we recall the Galilei velocity transformation formula $\mathbf{v}' = \mathbf{v} - \mathbf{u}$ and obtain $\mathcal{J}' = \mathbf{j}' - q'\mathbf{v}' = \mathbf{j} - q\mathbf{u} - q(\mathbf{v} - \mathbf{u}) = \mathbf{j} - q\mathbf{v} = \mathcal{J}$. Similarly, we have $\mathcal{H}' = \mathbf{H}' - \mathbf{v}' \times \mathbf{D}' = \mathbf{H} - \mathbf{u} \times \mathbf{D} - (\mathbf{v} - \mathbf{u}) \times \mathbf{D} = \mathbf{H} - \mathbf{v} \times \mathbf{D} = \mathcal{H}$.

We have decided to regard the integral laws (5.6) as the basic and general expression of the principle of charge conservation. Wherever the integrands are smooth ('between' elements of discontinuity), they lead to the local laws

$$\operatorname{div} \mathbf{D} = q,$$

$$\operatorname{curl} \mathcal{H} = \mathcal{J} + \overset{*}{\mathbf{D}}. \tag{5.8}$$

According to (5.3) (which can be read backwards), these are just Maxwell's equations (4.8), written in terms of Galilei-invariant quantities.

Let us now see what the integral laws give at a surface of discontinuity. In Figure 1, S is a surface of (possible) discontinuity which is moving with velocity $\mathbf{v}$ (not necessarily uniform over S). The pillbox has generators parallel to $\mathbf{n}$, a unit normal to the surface. We apply (5.6)$_1$ to the pillbox and assume that the surface carries a *surface charge density* (charge per unit area) σ. If we now pass to the limit in which the height of the pillbox vanishes, the contributions from space charges (if any) on either side of S will vanish, and we obtain

$$\mathbf{n} \cdot [\![\mathbf{D}]\!] = \sigma, \tag{5.9}$$

where the left hand side is calculated in accordance with the following prescription: having chosen $\mathbf{n}$ (in one of two possible ways), $[\![\mathbf{D}]\!] = \mathbf{D}_+ - \mathbf{D}_-$, where $\mathbf{D}_+$ is $\mathbf{D}$ on the side of S into which $\mathbf{n}$ is pointing, and $\mathbf{D}_-$ is $\mathbf{D}$ on the other side. This prescription ensures that (5.9) will hold

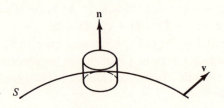

FIGURE 1

independently of the choice of $\mathbf{n}$: the alternative choice will change the signs of both $\mathbf{n}$ and $[\![\mathbf{D}]\!]$, and will therefore not affect their product. Of course, $\mathbf{D}_n$ will be continuous if, and only if, σ vanishes.

In Figure 2 we have a rectangle in a plane perpendicular to S. The orientation of the plane is defined by its normal $\mathbf{t}$, which is, of course, parallel to the surface. We apply $(5.6)_2$ to the rectangle and pass to the limit in which it collapses onto S. The flux $\int \mathbf{D} \cdot \mathbf{n} \, dS$ is equal to a sum of fluxes on either side of S. These vanish together with the width of the rectangle, and since $\mathbf{D}$ is smooth outside S, the time derivative of the flux, too, will vanish. Next, consider the current $\int \mathbf{j} \cdot \mathbf{t} \, dS$ passing through the rectangle as the latter collapses onto a segment Δl on S. We define the *surface current density* $\mathbf{K}$ in such a way that the amount of charge flowing through Δl is $\mathbf{K} \cdot \mathbf{t}\Delta l$. In the limit we then obtain

$$(\mathbf{t} \times \mathbf{n}) \cdot [\![\mathbf{H} - \mathbf{v} \times \mathbf{D}]\!] = (\mathbf{K} - \sigma\mathbf{v}) \cdot \mathbf{t},$$

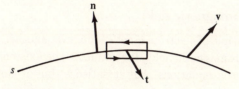

FIGURE 2

or, since **t** is arbitrary,

$$\mathbf{n} \times [\![\mathbf{H} - \mathbf{v} \times \mathbf{D}]\!] = \mathbf{K} - \sigma\mathbf{v}.$$

Using (5.9), this gives

$$\mathbf{n} \times [\![\mathbf{H}]\!] + v_n[\![\mathbf{D}]\!] = \mathbf{K}. \tag{5.10}$$

Here, too, the jumps are defined together with the choice of **n** in accordance with the prescription we have given following (5.9).

CHAPTER II

The Electromagnetic Field

and The Aether Relations

6. The electromagnetic field

The second principle of electromagnetism involves another primitive concept – the electromagnetic field. It postulates the existence of this field and lays down the equations that it satisfies, but it does not tell us what the field is supposed to do. In particular, it does *not* identify the electromagnetic field as a force field. That aspect, as we shall see, not only requires a separate postulate concerning the link between electromagnetism and mechanics (and thermodynamics); it also involves the constitutive properties of materials.

The second principle states: there exists an absolute, covariant, antisymmetric tensor $F_{\alpha\beta}$, called the *electromagnetic field*, that satisfies the equations

$$\epsilon^{\alpha\beta\gamma\delta}\frac{\partial F_{\gamma\delta}}{\partial x^\beta} = 0, \qquad \alpha = 1, 2, 3, 4. \tag{6.1}$$

Here $\epsilon^{\alpha\beta\gamma\delta}$ is the four-dimensional permutation symbol ($\epsilon^{1234} = 1$). We shall presently introduce another principle to connect the electromagnetic field F with the charge-current potential f. In the remainder of this section we discuss the equations (6.1) in a manner which is quite similar to the way in which we have discussed the equations (4.1). We shall therefore be brief.

PROBLEM

Problem 6–1. Prove that (6.1) are tensor equations.

18

Like the antisymmetric tensor f, F has six independent components, which we divide into two groups of three:

$$E_r = F_{r4}, \qquad \epsilon_{rst}B^t = F_{rs}, \qquad (6.2)$$

where ϵ_{rst} is numerically equal to ϵ^{rst}. Note the difference in the order in which the indices r and 4 appear in $(4.2)_1$ and $(6.2)_1$.

With these names for its components, F becomes

$$F = \begin{pmatrix} 0 & B^3 & -B^2 & E_1 \\ -B^3 & 0 & B^1 & E_2 \\ B^2 & -B^1 & 0 & E_3 \\ -E_1 & -E_2 & -E_3 & 0 \end{pmatrix}. \qquad (6.3)$$

The E_r are called the *components of the electric field*, and the B^r are called the *components of the magnetic field*. As in the discussion we have given of D^r and H_r in §4, we can justify referring to E_r and B^r as three-dimensional vectors.

The homogeneous system (6.1) is satisfied by setting

$$F_{\alpha\beta} = \frac{\partial V_\beta}{\partial x^\alpha} - \frac{\partial V_\alpha}{\partial x^\beta}, \qquad (6.4)$$

where V_α is an arbitrary four-dimensional covariant vector, called the *electromagnetic potential*. It is not uniquely determined by F.

If we use the notations (6.2) for the components of F, we can write (6.1) in the form

$$\mathrm{div}\, \mathbf{B} = 0,$$

$$\mathrm{curl}\, \mathbf{E} + \frac{\partial \mathbf{B}}{\partial t} = 0. \qquad (6.5)$$

This is the second pair of Maxwell's equations. Like the first pair (4.8) – and for a similar reason – these equations hold in all frames.

The dimensions of $\mathbf{E}$ and $\mathbf{B}$ are of course connected by $(6.5)_2$. If we denote the dimension of magnetic flux $\int B_n\, dS$ by Φ, the dimensions of $\mathbf{B}$ are Φ/L^2 and those of $\mathbf{E}$ are $\Phi/(LT)$.

The conventional way of introducing the electromagnetic potential is to note that (6.5) are solved by setting

$$\mathbf{B} = \mathbf{curl\ A},$$

$$\mathbf{E} = -\frac{\partial \mathbf{A}}{\partial t} - \mathbf{grad}\, V, \tag{6.6}$$

where the *vector potential* $\mathbf{A}$ and the *scalar potential* V are arbitrary. Of course, (6.6) is the same as (6.4) if we recognize that $(V_\alpha) = (\mathbf{A}, -V)$. The electromagnetic potential can be replaced by another one through the so-called *gauge transformation*

$$\tilde{\mathbf{A}} = \mathbf{A} + \mathbf{grad}\,\chi, \qquad \tilde{V} = V - \frac{\partial \chi}{\partial t}, \tag{6.7}$$

where χ is an arbitrary function. The new $\tilde{\mathbf{A}}$ and $\tilde{V}$ lead, through (6.6), to the same $\mathbf{E}$ and $\mathbf{B}$ as the old $\mathbf{A}$ and V. In four-dimensional language, the gauge transformation is $\tilde{V}_\alpha = V_\alpha + \partial \chi / \partial x^\alpha$.

In applications of electromagnetism we shall always use Maxwell's equations (6.5), rather than the tensor equations (6.1). Correspondingly, we shall always use (6.6) and (6.7), rather than (6.4) and $\tilde{V}_\alpha = V_\alpha + \partial \chi / \partial x^\alpha$. But (6.5)–(6.7) cannot tell us how the fields $\mathbf{E}$ and $\mathbf{B}$ in one frame are connected with their values (at the same event) in another one. In order to determine this, we must return to the definitions (6.2) of $\mathbf{E}$ and $\mathbf{B}$ and use the fact that F is a tensor.

We have assumed that F is an absolute, covariant tensor. If we subject it to the Galilei transformation (2.7) and express the result in terms of $\mathbf{E}$ and $\mathbf{B}$, we obtain

$$\mathbf{B}' = \mathbf{B},$$

$$\mathbf{E}' = \mathbf{E} + \mathbf{u} \times \mathbf{B}, \tag{6.8}$$

where there is now a change of sign in the *first* formula whenever $\det A = -1$. As in §5, we can integrate Maxwell's equations (6.5) to obtain the integral formulae

$$\oint B_n \, dS = 0,$$

$$\oint E_s \, ds = -\frac{d}{dt} \int B_n \, dS, \qquad (6.9)$$

where, in the second formula, the open surface is fixed. Again, as in §5, if $\mathbf{v}$ is any velocity field,

$$\mathbf{curl}\,(\mathbf{E} + \mathbf{v} \times \mathbf{B}) = -\frac{\partial \mathbf{B}}{\partial t} + \mathbf{curl}\,(\mathbf{v} \times \mathbf{B})$$

$$= -\left[\frac{\partial \mathbf{B}}{\partial t} + \mathbf{v}\,\mathrm{div}\,\mathbf{B} - \mathbf{curl}\,(\mathbf{v} \times \mathbf{B})\right] = -\overset{*}{\mathbf{B}}, \qquad (6.10)$$

where use has been made of $(6.5)_1$ and of the moving flux derivative (5.4). From this we obtain, for a moving surface,

$$\oint (\mathbf{E} + \mathbf{v} \times \mathbf{B})_s \, ds = -\frac{d}{dt} \int B_n \, dS. \qquad (6.11)$$

This is the general formulation of Faraday's law of electromagnetic induction. We note, as in §5, that the integrands are Galilei-invariants. The invariant

$$\mathcal{E} = \mathbf{E} + \mathbf{v} \times \mathbf{B} \qquad (6.12)$$

will be called the *electromotive intensity*.

As in §5, we shall now regard $(6.9)_1$ and (6.11) as the basic laws that express the second principle of electromagnetism for piecewise smooth fields. Wherever the fields are smooth, these laws lead to the equations

$$\mathrm{div}\,\mathbf{B} = 0,$$

$$\mathbf{curl}\,\mathcal{E} = -\overset{*}{\mathbf{B}}, \qquad (6.13)$$

which, according to (6.10), are the same as the second pair (6.5) of Maxwell's equations. On surfaces of discontinuity, they lead to the jump conditions

$$\mathbf{n} \cdot [\![\mathbf{B}]\!] = 0,$$

$$\mathbf{n} \times [\![\mathbf{E}]\!] - v_n [\![\mathbf{B}]\!] = 0. \qquad (6.14)$$

PROBLEMS

Problem 6–2. The dual of F is defined by

$$(\text{dual } F)^{\alpha\beta} = \frac{1}{2}\epsilon^{\alpha\beta\gamma\delta}F_{\gamma\delta}.$$

The equations (6.1) can therefore be written as

$$\frac{\partial(\text{dual } F)^{\alpha\beta}}{\partial x^\beta} = 0.$$

Show that

$$\frac{1}{4}(\text{dual}F)^{\alpha\beta}F_{\alpha\beta} = B^r E_r.$$

Problem 6–3. Prove that

$$\frac{1}{2}f^{\alpha\beta}F_{\alpha\beta} = B^r H_r - D^r E_r.$$

This is a scalar density.

Problem 6–4. Let $f^{\alpha\beta}$ be an *arbitrary* antisymmetric tensor density, and σ^α an arbitrary vector density. If the scalar density $\sigma^\alpha V_\alpha - \frac{1}{2}f^{\alpha\beta}F_{\alpha\beta}$ is integrated over a four-dimensional volume, the result is the invariant

$$A = \int \left(\sigma^\alpha V_\alpha - \frac{1}{2}f^{\alpha\beta}F_{\alpha\beta}\right) d^4x.$$

If, in the integrand, F is expressed in terms of V_α by (6.4) – which is equivalent to assuming the second pair (6.5) of Maxwell's equations – A becomes a functional $A[V_\alpha]$ of the components of V. Prove that, if the V_α have fixed values on the boundary, but are otherwise arbitrary, the first pair (4.8) of Maxwell's equations results from the requirement that $A[V_\alpha]$ be stationary.

7. The Maxwell-Lorenz aether relations

Maxwell's equations,

$$\sigma^\alpha = \frac{\partial f^{\alpha\beta}}{\partial x^\beta}, \qquad \epsilon^{\alpha\beta\gamma\delta}\frac{\partial F_{\gamma\delta}}{\partial x^\beta} = 0, \qquad (7.1)$$

or, in three-dimensional language (we use the space-saving notation $f_t = \partial f/\partial t$),

$$\text{div } \mathbf{D} = q,$$

$$\text{curl } \mathbf{H} - \mathbf{D}_t = \mathbf{j},$$

$$\text{div } \mathbf{B} = 0,$$

$$\text{curl } \mathbf{E} + \mathbf{B}_t = 0, \tag{7.2}$$

are tensor equations. They are therefore invariant with respect to *all* transformations of space-time (not just Galilei transformations). Obviously then, the claim, often made, that they are invariant only with respect to a *special* transformation – the Lorenz transformation – cannot be true. This claim is really connected with the third principle of electromagnetism, which we are about to introduce. Its aim is threefold: (a) to remove the arbitrariness in the charge-current potential $f = \{\mathbf{H}, \mathbf{D}\}$; (b) to connect the hitherto independent charge-current potential f and the electromagnetic field F; (c) to create a link between electromagnetism and mechanics.

The third principle states: there exists an inertial frame in which the relations

$$\mathbf{D} = \epsilon_0 \mathbf{E}, \qquad \mathbf{H} = \mathbf{B}/\mu_0, \tag{7.3}$$

with ϵ_0 and μ_0 two positive, universal constants, hold everywhere and at all times – inside material bodies as well as in the empty space between them. We emphasize that the charge-current potentials $\mathbf{D}$ and $\mathbf{H}$ in (7.3) refer to the *total* charges and currents, including those which we shall later associate with polarization and magnetization.

We recall that an inertial frame is one in which the laws of mechanics hold (*without* the addition of 'fictitious' forces and torques). According to classical mechanics, all inertial frames follow from any one of them by applying Galilei transformations. As we have seen, these transformations have the following effects on $\mathbf{D}$, $\mathbf{H}$, $\mathbf{E}$ and $\mathbf{B}$:

$$\mathbf{D}' = \mathbf{D}, \qquad \mathbf{H}' = \mathbf{H} - \mathbf{u} \times \mathbf{D},$$

$$\mathbf{E}' = \mathbf{E} + \mathbf{u} \times \mathbf{B}, \qquad \mathbf{B}' = \mathbf{B}. \tag{7.4}$$

Clearly, the relations (7.3) cannot hold in all of the classically inertial frames. On the contrary, the frame in which they do hold emerges as a special, preferred one. It is called *the aether frame*, and the relations

(7.3) are called *the aether relations*. We shall presently comment on the reasons behind these names.

If we recall the dimensions of $\mathbf{D}$ and $\mathbf{E}$, we see that the dimensions of the universal constant ϵ_0 are $(C/\Phi)/(L/T)$. Similarly, the dimensions of μ_0 are $(\Phi/C)/(L/T)$. The dimensions of $(\epsilon_0\mu_0)^{-1}$ are therefore the dimensions of velocity squared. We now show that this is the velocity of an electromagnetic disturbance in vacuum.

The jump conditions across a surface of discontinuity in vacuum ($\mathbf{K} = \sigma = 0$) which is moving with velocity $\mathbf{v}$ are

$$\mathbf{n} \cdot [\![\mathbf{D}]\!] = 0,$$

$$\mathbf{n} \times [\![\mathbf{H}]\!] + v_n[\![\mathbf{D}]\!] = 0,$$

$$\mathbf{n} \cdot [\![\mathbf{B}]\!] = 0,$$

$$\mathbf{n} \times [\![\mathbf{E}]\!] - v_n[\![\mathbf{B}]\!] = 0. \tag{7.5}$$

In the aether frame, we also have the relations (7.3). It is clear that, if $v_n = 0$, $\mathbf{E}$ and $\mathbf{B}$ (and, by the aether relations, $\mathbf{D}$ and $\mathbf{H}$ as well) are all continuous. A surface across which these fields are discontinuous – it is called an *electromagnetic shock* – must therefore be moving. According to $(7.5)_2$, $[\![\mathbf{D}]\!]$ is perpendicular to $\mathbf{n}$, so that $(7.5)_1$ is a consequence of $(7.5)_2$. Similarly, $(7.5)_3$ follows from $(7.5)_4$. It is therefore sufficient to consider the two conditions

$$\mathbf{n} \times [\![\mathbf{B}]\!]/\mu_0 + \epsilon_0 v_n[\![\mathbf{E}]\!] = 0,$$

$$\mathbf{n} \times [\![\mathbf{E}]\!] - v_n[\![\mathbf{B}]\!] = 0. \tag{7.6}$$

According to (7.6), the jumps $[\![\mathbf{E}]\!]$ and $[\![\mathbf{B}]\!]$ are parallel to the surface and perpendicular to each other. If we eliminate one of the jumps between the two equations, we obtain for v_n the condition

$$v_n^2 = (\epsilon_0\mu_0)^{-1}. \tag{7.7}$$

Since electromagnetism associates propagating electromagnetic disturbances with light, we identify this velocity with the speed of light c:

$$c^2 = (\epsilon_0\mu_0)^{-1}. \tag{7.8}$$

The speed of light has been measured with continually increasing accuracy over a long time. An approximate value of c in terms of the old platinum metre in Paris is 3×10^8 m/s. Since ϵ_0 and μ_0 are *universal constants*, we can also *assume* a value of c, together with a given unit of time, in order to fix the unit of length. Several international committees on physical data and constants have in fact recommended that we set

$$c = 299\,792\,458\,\text{m/s} \tag{7.9}$$

as the definition of the metre in terms of the second. By definition, then, the value of c is *exactly* the one given by (7.9), and any improved measurement of the speed of light is really an improved measurement of the metre.

The founders of electromagnetism believed that propagating disturbances required the presence of a medium. But, as the foregoing discussion has shown, an electromagnetic shock can propagate inside, as well as outside, material bodies, the only condition being the absence of charges and currents. This has led to the idea of a medium, called the *aether*, which pervades all matter and fills the empty space between material bodies, but has none of the properties of ordinary matter – no mass, no charge, no temperature, etc. Its only role is to provide a seat for electromagnetic phenomena, and the aether frame is the one in which the aether is at rest. In other frames, particularly in other inertial frames, we should be able, by a simultaneous application of the laws of mechanics and electromagnetism, to detect an *aether wind*, that is, a motion with respect to the aether. This means, for example, that if we measure the speed of light in a frame that is not the aether frame, we should find a value different from c, a result of the Galilei composition of the velocity c in the aether frame and the aether wind. Of course there may be other methods of detecting an aether wind; light is not the only phenomenon that follows from electromagnetic theory.

8. Lorenz transformations

It is well known that all attempts, beginning about a century ago, to detect an aether wind have failed. In particular, the velocity of light has turned out to be c in different frames which were obviously moving relative to each other. Perhaps, then, there is more than one aether frame. This question can be answered by the theory developed so far.

Let $(x) = (\mathbf{x}, t)$ stand for the Cartesian coordinates and time in one (or *the*) aether frame. Let $g_{\alpha\beta}$ be the absolute, symmetric, covariant tensor that, in this frame, has the components

$$g = \begin{pmatrix} \delta_{rs} & 0 \\ 0 & -c^2 \end{pmatrix}, \tag{8.1}$$

where δ is the Kronecker symbol. The components of $g_{\alpha\beta}$ in any other frame will be determined by the tensor transformation rule; hence (8.1) defines $g_{\alpha\beta}$ in any frame.† The determinant of $g_{\alpha\beta}$ is $-c^2$ in the frame $(\mathbf{x})$, and can be shown to be a scalar density of weight 2. Let $g^{\alpha\beta}$ denote the contravariant tensor that is the reciprocal of $g_{\alpha\beta}$, defined by

$$g_{\alpha\gamma}g^{\gamma\beta} = \delta_\alpha^\beta. \tag{8.2}$$

In the frame (x), $g^{\alpha\beta}$ has the same components as $g_{\alpha\beta}$, except for $g^{44} = -c^{-2} = -\epsilon_0\mu_0$.

Consider now the tensor equation

$$f^{\alpha\beta} = \sqrt{\frac{\epsilon_0}{\mu_0}} (-\det g_{\mu\nu})^{\frac{1}{2}} g^{\alpha\gamma} g^{\beta\delta} F_{\gamma\delta}. \tag{8.3}$$

This is a tensor equation, because both sides agree in rank and weight. It is true in the aether frame (x), because there (8.3) is the same as the aether relations (7.3), as can easily be verified. We can therefore take (8.3) to be the tensor expression of the aether relations.

Now, if there are any other aether frames (x'), in which (7.3) hold, they must be those in which the components of

$$g^{\alpha'\beta'}(x') = \frac{\partial x^{\alpha'}}{\partial x^\alpha} \frac{\partial x^{\beta'}}{\partial x^\beta} g^{\alpha\beta}(x) \tag{8.4}$$

are equal to the constant components of $g^{\alpha\beta}(x)$ in (x). It is clear that the transformations leading to these frames must be linear. They can

---

† The electromagnetic tensor $g_{\alpha\beta}$, with its definition involving $c^2 = (\epsilon_0\mu_0)^{-1}$, is the fundamental tensor of Einstein's theory of gravitation (the general theory of relativity).

all be obtained from the special transformation

$$x' = \frac{x - ut}{\sqrt{1 - u^2/c^2}}, \quad y' = y, \quad z' = z, \quad t' = \frac{t - ux/c^2}{\sqrt{1 - u^2/c^2}}, \qquad (8.5)$$

by orthogonal transformations of the space coordinates and time inversions. Their general form is

$$x^{r'} = A^{r'}_r \{[\delta^r_s + (\gamma - 1)u^{-2}u^r u_s]x^s - \gamma u^r x^4\},$$

$$x^{4'} = \pm\gamma(x^4 - u_r x^r/c^2), \qquad (8.6)$$

where A is an orthogonal matrix, $\gamma = (1 - u^2/c^2)^{-1/2}$, $u_r = u^r$ and $u^2 = u_r u^r$ must be less than c^2 for the transformation to be real. The transformations (8.6) are called *Lorenz transformations*, and the frames that are obtained by them from the aether frame are called *Lorenz frames*. Of course, they are all aether frames.

The theory of relativity identifies the Lorenz frames with the inertial frames and replaces the Galilei transformations with the Lorenz transformations. This requires a major revision of all the laws of physics, except, of course, those of electromagnetism. Thus mechanics, thermodynamics, the theory of gravitation and quantum mechanics all have to be revised in order to conform to the new transformation laws. This revision is still underway. Since, in the new theory, all inertial frames are aether frames, the aether (which would have to be at rest in each of the Lorenz frames) has gone out of vogue and is never mentioned in the theory of relativity.

PROBLEM

Problem 8–1. Derive the following transformation formulae from the transformation laws of the tensors $f^{\alpha\beta}$ and $F_{\alpha\beta}$ under proper Lorenz transformations ($\det A = 1$, + sign in (8.6)$_2$):

$$\mathbf{E}' = \gamma(\mathbf{E} + \mathbf{u} \times \mathbf{B}) - \frac{\gamma^2}{\gamma + 1}\mathbf{u}(\mathbf{u} \cdot \mathbf{E})/c^2,$$

$$\mathbf{B}' = \gamma(\mathbf{B} - \mathbf{u} \times \mathbf{E}/c^2) - \frac{\gamma^2}{\gamma + 1}\mathbf{u}(\mathbf{u} \cdot \mathbf{B})/c^2,$$

$$\mathbf{D}' = \gamma(\mathbf{D} + \mathbf{u} \times \mathbf{H}/c^2) - \frac{\gamma^2}{\gamma+1}\mathbf{u}(\mathbf{u} \cdot \mathbf{D})/c^2,$$

$$\mathbf{H}' = \gamma(\mathbf{H} - \mathbf{u} \times \mathbf{D}) - \frac{\gamma^2}{\gamma+1}\mathbf{u}(\mathbf{u} \cdot \mathbf{H})/c^2,$$

$$\mathbf{j}' = \mathbf{j} + [(\gamma - 1)(\mathbf{u} \cdot \mathbf{j})/u^2 - \gamma q]\mathbf{u},$$

$$q' = \gamma[q - (\mathbf{u} \cdot \mathbf{j})/c^2]. \tag{8.7}$$

In an aether frame, Maxwell's equations can be written in the form

$$\operatorname{div} \epsilon_0 \mathbf{E} = q,$$

$$\operatorname{curl} \mathbf{B}/\mu_0 - \epsilon_0 \mathbf{E}_t = \mathbf{j},$$

$$\operatorname{div} \mathbf{B} = 0,$$

$$\operatorname{curl} \mathbf{E} + \mathbf{B}_t = 0. \tag{8.8}$$

These are often regarded as linear equations for the electromagnetic field with q and $\mathbf{j}$ playing the part of sources. But this is only true in situations which are very special – like the electrostatics of conductors – or so simple as to be degenerate – point charges, linear currents and the like – because the 'sources' q and $\mathbf{j}$ are usually not given. In fact, it is normal for the charges and currents to depend in a complicated manner on the electromagnetic field itself. The equations may then become non-linear, and may even cease to be partial differential equations.

According to the aether relations, the components $\mathbf{H}$ and $\mathbf{D}$ of the charge-current potential are, respectively, proportional to $\mathbf{B}$ and $\mathbf{E}$. There is no harm in calling $\mathbf{H}$ 'the magnetic field $\mathbf{H}$', and $\mathbf{D}$ 'the electric field $\mathbf{D}$', so long as we remember, *especially* in frames that are not aether frames, that $\mathbf{H}$ and $\mathbf{D}$ are really components of the charge-current potential.†

---

† In the old literature, $\mathbf{H}$ is called the magnetic field or the magnetic intensity, and $\mathbf{B}$ the magnetic induction.

Of the three principles of electromagnetism, the third is the easiest to forget. Most students of electromagnetism have no difficulty in memorizing the two pairs of Maxwell's equations, but many of them are likely to manufacture 'paradoxes' by thoughtless application of (8.8) to frames that are not aether frames.

We conclude with an example. In an aether frame O, described by the cylindrical coordinates (r, ϕ, z), let the charge-current potential have the components

$$\mathbf{D} = (0, 0, 0), \qquad \mathbf{H} = (0, i/(2\pi r), 0). \tag{8.9}$$

Obviously, the charge density vanishes everywhere. It is easy to verify that the current density also vanishes, except along the z axis; and that a linear current i flows along this axis in the positive direction. The aether relations determine the electromagnetic field:

$$\mathbf{E} = (0, 0, 0), \qquad \mathbf{B} = (0, \mu_0 i/(2\pi r), 0). \tag{8.10}$$

Let O' now be a frame that is moving, relative to O, with velocity u, parallel to the z axis in the direction of the current. In order to obtain the various quantities in the frame O', according to classical physics, we apply a Galilei transformation. From (7.4) we obtain

$$\mathbf{D}' = \mathbf{D}, \qquad \mathbf{H}' = \mathbf{H},$$

$$\mathbf{E}' = (-\mu_0 u i/(2\pi r), 0, 0), \qquad \mathbf{B}' = \mathbf{B}. \tag{8.11}$$

Since the charge-current potential is the same as in O, the charge and current distributions are also the same: no charge anywhere, and a linear current i flowing along the z axis. The magnetic field $\mathbf{B}$ is also the same as in O, but there is an electric field directed towards the z axis. If there is no charge anywhere, where does this electric field come from? This question arises because the third principle has been forgotten. The electric field 'comes from charge' only in an aether frame, through the relation $\mathbf{E} = \mathbf{D}/\epsilon_0$. The frame O' is not an aether frame, hence $\mathbf{E}' \neq \mathbf{D}'/\epsilon_0 = 0$, and there is no contradiction between the absence of charge and a non-zero electric field.

The relativistic treatment of this particular example is trouble-free. The frame O', now obtained from O by a Lorenz transformation, is an aether frame as well. Instead of (8.11), we have

$$\mathbf{D}' = \left(-\gamma ui/(2\pi rc^2), 0, 0\right), \qquad \mathbf{H}' = \mathbf{H},$$

$$\mathbf{E}' = \left(-\gamma \mu_0 ui/(2\pi r), 0, 0\right), \qquad \mathbf{B}' = \mathbf{B}, \tag{8.12}$$

with $\gamma = (1 - u^2/c^2)^{-1/2}$. Now there is also a negative line charge, of amount $\gamma ui/c^2$ per unit length, on the z axis. It would now be correct to say that $\mathbf{E}'$, which differs from the classical $\mathbf{E}'$ of $(8.11)_3$ only by the factor γ, 'comes from' this line charge, because O' is now an aether frame.

It is important to note that we consider the relativistic treatment to be superior to the classical one – to which it reduces in the limit of small velocities – because of its evident successes when applied to experiments. But neither treatment is superior to the other from a purely logical point of view, because both are consistent logical structures. Moreover, we shall see later on that there are cases in which a non-relativistic theory cannot be merely regarded as an adequate approximation, valid at low speeds, because a corresponding relativistic theory does not exist. It therefore pays to practise thinking in terms of either description, as if the two were alternatives of equal standing.

PROBLEM

Problem 8–2. The scalar density

$$\mathcal{L} = \sigma^\alpha V_\alpha - \frac{1}{2} f^{\alpha\beta} F_{\alpha\beta}$$

may be regarded as a Lagrangian density, leading to an invariant action $S = \int \mathcal{L} \, d^4x$ (cf. Problem 6–4). Show that, in terms of the fields and the four-dimensional volume in the aether frame,

$$S = \int \left(\frac{\epsilon_0 E^2}{2} - \frac{B^2}{2\mu_0} + \mathbf{A} \cdot \mathbf{j} - Vq \right) d^4x.$$

CHAPTER III

Polarization

and Magnetization

9. Material response functions

Experience shows that many materials *respond*, as it were, to an electromagnetic field by setting up charge and current distributions. This takes place in a myriad of forms, depending on the kind of material, on its state, and even on its history. Indeed, the resulting charge and current distributions are so diverse that, in constructing a theory of these phenomena, it is wise to keep an open mind. In particular, we should not impose on them any restrictions beyond those that we regard as absolutely necessary. Such a restriction is the principle of charge conservation, which we are certainly not about to renounce. A simple way of imposing it is to use what we have already learned and characterize the charge and current distribution of material response by a charge-current potential,

$$f_R = \{\mathbf{M}, -\mathbf{P}\}, \tag{9.1}$$

where $\mathbf{M}$ and $\mathbf{P}$ (the sign in (9.1) is conventional) are two piecewise smooth vector fields which depend in the most general way on the material, on its state – such as its density, temperature, elastic strain – and on the electromagnetic field, that is, $\mathbf{E}$ and $\mathbf{B}$. Indeed, $\mathbf{M}$ and $\mathbf{P}$ may at any time depend, not only on the values of these variables at that time, but even on their histories. The field $\mathbf{M}$ is called the *magnetization*; the field $\mathbf{P}$, the *dielectric polarization*, or simply the *polarization*.

The corresponding charge and current distributions – often called *bound* charges and currents – can now be obtained from the relations of

31

§4–5. Wherever $\mathbf{M}$ and $\mathbf{P}$ are smooth,

$$q_R = -\operatorname{div}\mathbf{P}, \qquad \mathbf{j}_R = \operatorname{\mathbf{curl}}\mathbf{M} + \mathbf{P}_t \qquad (9.2)$$

(cf. (4.6)–(4.8)). These q_R and $\mathbf{j}_R$ obey the law of charge conservation (1.7), as they must. On a surface across which $\mathbf{M}$ or $\mathbf{P}$ suffer discontinuities – we expect this to happen, for example, on the boundary between two materials – there will be surface charges and currents (cf. (5.9)–(5.10)):

$$\sigma_R = -\mathbf{n}\cdot[\![\mathbf{P}]\!], \qquad \mathbf{K}_R = \mathbf{n}\times[\![\mathbf{M}]\!] - v_n[\![\mathbf{P}]\!]. \qquad (9.3)$$

Under Galilei transformations, $\mathbf{P}$ and $\mathbf{M}$ will transform as follows:

$$\mathbf{P}' = \mathbf{P},$$

$$\mathbf{M}' = \mathbf{M} + \mathbf{u}\times\mathbf{P} \qquad (9.4)$$

(cf. (4.5)). Thus $\mathbf{P}$ is a Galilei-invariant, but $\mathbf{M}$ is not.

PROBLEM

Problem 9–1. Find the transformation formulae of $\mathbf{P}$ and $\mathbf{M}$ under a Lorenz transformation.

As in §5, we can construct two other fields that are Galilei-invariant (cf. (5.7)):

$$\mathcal{M} = \mathbf{M} + \mathbf{v}\times\mathbf{P},$$

$$\mathcal{J}_R = \mathbf{j}_R - q_R\mathbf{v} = \operatorname{\mathbf{curl}}\mathbf{M} + \mathbf{P}_t + \mathbf{v}\operatorname{div}\mathbf{P} = \operatorname{\mathbf{curl}}\mathcal{M} + \overset{*}{\mathbf{P}}. \qquad (9.5)$$

The first invariant, $\mathcal{M}$, is called the *Lorenz magnetization*.

We shall now discuss some simple properties of the foregoing charge-current distributions. Consider, first, an isolated body (in empty space) which is polarized, but not magnetized, i.e. $\mathbf{M} = 0$. It then has a volume charge density $-\operatorname{div}\mathbf{P}$ and (if we choose $\mathbf{n}$ to point outward) a surface charge density $-\mathbf{n}\cdot[\![\mathbf{P}]\!] = \mathbf{n}\cdot\mathbf{P} = P_n$, since $\mathbf{P} = 0$ in the empty space outside the body. Taken together, the total charge on the body is $\int -\operatorname{div}\mathbf{P}\, dV + \oint P_n\, dS = 0$, by Gauss's theorem.

PROBLEM

Problem 9–2. Prove that, for an isolated, polarized, body,

$$\int (\mathbf{x} - \mathbf{x}_O) q_R \, dV + \oint (\mathbf{x} - \mathbf{x}_O) \sigma_R \, dS = \int \mathbf{P} \, dV, \qquad (9.6)$$

where $\mathbf{x}_O$ is the position of a fixed point O and $\mathbf{x}$ is the position at which the integrand is evaluated.

The expression on the left hand side of (9.6) is called the *dipole moment* with respect to O of the charge distribution. Equation (9.6) shows that the dipole moment does not depend on O (this is because the total charge vanishes), and that $\mathbf{P}$ is the dipole moment per unit volume.

We can now similarly discuss some analogous properties of isolated, magnetized, bodies: $\mathbf{M} \neq 0$, but $\mathbf{P} = 0$. In such a body, there is a current density $\mathbf{j}_R = \mathbf{curl}\,\mathbf{M}$ and (again choosing $\mathbf{n}$ to point outward) a surface current density $\mathbf{K}_R = \mathbf{n} \times [\![\mathbf{M}]\!] = \mathbf{M} \times \mathbf{n}$.

PROBLEMS

Problem 9–3. An isolated, magnetized, body is bisected by a surface S. Use Stokes's theorem to prove that the net total magnetization current through S vanishes.

Problem 9–4. Prove that, in an isolated, magnetized, body (with the notation of Problem 9–2),

$$\tfrac{1}{2} \int (\mathbf{x} - \mathbf{x}_O) \times \mathbf{j}_R \, dV + \tfrac{1}{2} \oint (\mathbf{x} - \mathbf{x}_O) \times \mathbf{K}_R \, dS = \int \mathbf{M} \, dV. \qquad (9.7)$$

The expression on the left hand side of (9.7) is called the *magnetic moment* with respect to O of the current distribution. Equation (9.7) shows that the magnetic moment is independent of O, and that $\mathbf{M}$ is the magnetic moment per unit volume.

The foregoing properties of polarization and magnetization in isolated bodies are often used to *define* $\mathbf{P}$ and $\mathbf{M}$ on the basis of the atomic constitution of matter. This requires a rather involved process of averaging over microscopic charges and currents, with one or two infinite series of 'higher-order multipole moments' cast away as one moves off from the isolated body, presumably in the direction of a measuring apparatus. Quite apart from the fact that polarization and magnetization emerge as quantities that are only approximations, one is left wondering whether any quantity can ever be averaged in the absence of complete information regarding its distribution, whether electrons and nuclei can be assumed to have definite (even if unknown) positions *and* momenta, in defiance of the uncertainty principle, whether $(9.3)_2$ can be correctly obtained in this way, and so on. There are also the usual danger signs that one has come to expect in theories of this kind – philosophical reflections on the imperfection inherent in all physical quantities, the finite accuracy of measurements, etc. In any case, we know that all this is unnecessary, because the definition of $\mathbf{P}$ and $\mathbf{M}$ follows quite naturally from the principle of charge conservation. The atomic constitution of matter is as irrelevant here as the observation, in mechanics, that a rigid body, like the physical pendulum, 'really' consists of many atoms. How many atoms are there in a *mathematical* pendulum, and what difference does it make?

10. The partial potentials; Maxwell's equations in media

The bound charges and currents that arise from the response of a material to an electromagnetic field do not necessarily constitute the total charge residing in the material or the total current passing through it. If there are other charges and currents, besides the bound ones, we call them *free*, and designate them as q_F and $\mathbf{j}_F$:

$$q_F = q - q_R, \qquad \mathbf{j}_F = \mathbf{j} - \mathbf{j}_R. \tag{10.1}$$

Correspondingly, we define the *partial charge-current potential* as

$$f_F = \{\mathbf{H}_F, \mathbf{D}_F\} = f - f_R = \{\mathbf{H} - \mathbf{M}, \mathbf{D} + \mathbf{P}\}. \tag{10.2}$$

In terms of these, the first pair of Maxwell's equations becomes

$$\operatorname{div} \mathbf{D}_F = q_F,$$

$$\mathbf{curl}\, \mathbf{H}_F - \frac{\partial \mathbf{D}_F}{\partial t} = \mathbf{j}_F. \tag{10.3}$$

The jump conditions, too, can be obtained by this process of subtraction:

$$\sigma_F = \mathbf{n} \cdot [\![\mathbf{D}_F]\!],$$

$$\mathbf{K}_F = \mathbf{n} \times [\![\mathbf{H}_F]\!] + v_n[\![\mathbf{D}_F]\!]. \tag{10.4}$$

Alongside these, we must now add the second pair of Maxwell's equations and the aether relations. If we substitute the latter in (10.2), we obtain $\mathbf{H}_F = \mathbf{H} - \mathbf{M} = \mathbf{B}/\mu_0 - \mathbf{M}$, $\mathbf{D}_F = \mathbf{D} + \mathbf{P} = \epsilon_0\mathbf{E} + \mathbf{P}$. (We recall that the aether relations refer to the *total* potentials $\mathbf{H}$ and $\mathbf{D}$, not to the *partial* potentials $\mathbf{H}_F$ and $\mathbf{D}_F$.) Thus, in an aether frame,

$$\mathbf{D}_F = \epsilon_0\mathbf{E} + \mathbf{P},$$

$$\mathbf{H}_F = \mathbf{B}/\mu_0 - \mathbf{M}. \tag{10.5}$$

A glance at (10.3)–(10.5) shows that the total charge-current $(\mathbf{j}, q)$ and the total charge-current potential $\{\mathbf{H}, \mathbf{D}\}$ have disappeared, and the equations contain only the partial, or free, quantities. We may therefore leave out the subscript $_F$ and rewrite all the equations:

$$\operatorname{div} \mathbf{D} = q,$$

$$\operatorname{curl} \mathbf{H} = \mathbf{j} + \mathbf{D}_t,$$

$$\operatorname{div} \mathbf{B} = 0,$$

$$\operatorname{curl} \mathbf{E} = -\mathbf{B}_t,$$

$$\mathbf{D} = \epsilon_0\mathbf{E} + \mathbf{P},$$

$$\mathbf{H} = \mathbf{B}/\mu_0 - \mathbf{M}. \tag{10.6}$$

The first two equations $(10.6)_{1,2}$ *look* like (4.8), but they are not the same, because in (10.6) q and $\mathbf{j}$ are the free charge and current densities, and $\mathbf{D}$ and $\mathbf{H}$ the corresponding partial potentials. That is, of course, why $\mathbf{P}$ and $\mathbf{M}$ now appear in the aether relations $(10.6)_{5,6}$. The partial potential $\mathbf{D}$ is called the *electric displacement*, or the *displacement*.

Of course, we can also leave out the subscript in (10.4). The complete list of jump conditions becomes

$$\mathbf{n} \cdot [\![\mathbf{D}]\!] = \sigma,$$

$$\mathbf{n} \times [\![\mathbf{H}]\!] + v_n [\![\mathbf{D}]\!] = \mathbf{K},$$

$$\mathbf{n} \cdot [\![\mathbf{B}]\!] = 0,$$

$$\mathbf{n} \times [\![\mathbf{E}]\!] - v_n [\![\mathbf{B}]\!] = 0. \tag{10.7}$$

Again, σ and $\mathbf{K}$ are now free surface charge and surface current densities, and $\mathbf{D}$ and $\mathbf{H}$ are related to $\mathbf{E}$ and $\mathbf{B}$ through $(10.6)_{5,6}$.

We can again form Galilei invariants:

$$\boldsymbol{\mathcal{J}} = \mathbf{j} - q\mathbf{v}, \qquad \boldsymbol{\mathcal{H}} = \mathbf{H} - \mathbf{v} \times \mathbf{D}. \tag{10.8}$$

The first, $\boldsymbol{\mathcal{J}}$, is the (free) *conduction current density*. The second, $\boldsymbol{\mathcal{H}}$, is called the *magnetomotive intensity*.

PROBLEM

Problem 10–1. Prove that, in an aether frame, the magnetomotive intensity is

$$\boldsymbol{\mathcal{H}} = \mathbf{B}/\mu_0 - \epsilon_0 \mathbf{v} \times \mathbf{E} - \boldsymbol{\mathcal{M}}, \tag{10.9}$$

where $\boldsymbol{\mathcal{M}}$ is the Lorenz magnetization.

In terms of these Galilei invariants and the electromotive intensity $\boldsymbol{\mathcal{E}} = \mathbf{E} + \mathbf{v} \times \mathbf{B}$, equations (10.6) take the form

$$\operatorname{div} \mathbf{D} = q,$$

$$\operatorname{curl} \boldsymbol{\mathcal{H}} = \boldsymbol{\mathcal{J}} + \overset{*}{\mathbf{D}},$$

$$\operatorname{div} \mathbf{B} = 0,$$

$$\operatorname{curl} \boldsymbol{\mathcal{E}} = -\overset{*}{\mathbf{B}},$$

$$\mathbf{D} = \epsilon_0 \mathbf{E} + \mathbf{P},$$

$$\boldsymbol{\mathcal{H}} = \mathbf{B}/\mu_0 - \epsilon_0 \mathbf{v} \times \mathbf{E} - \boldsymbol{\mathcal{M}} \tag{10.10}$$

We shall make extensive use of the systems (10.6) or (10.10) and the jump conditions (10.7).

As an example, we consider the dragging of light by a dielectric.† A classical (i.e. non-relativistic), linear dielectric is defined by the response functions

$$\mathbf{P} = \epsilon_0 \chi \boldsymbol{\mathcal{E}}, \qquad \boldsymbol{\mathcal{M}} = 0, \tag{10.11}$$

where χ, the *dielectric susceptibility*, is a positive function which is independent of the electromagnetic field, but may depend on such Galilei invariants as the density or the temperature. The definition (10.11) of the response functions is therefore Galilei-invariant. The aether relations for the dielectric follow from $(10.6)_{5,6}$ and $(9.5)_1$:

$$\mathbf{D} = \epsilon_0 \left[\mathbf{E} + \chi (\mathbf{E} + \mathbf{v} \times \mathbf{B}) \right],$$

$$\mathbf{H} = \mathbf{B}/\mu_0 + \mathbf{v} \times \mathbf{P} = \mathbf{B}/\mu_0 + \epsilon_0 \chi \mathbf{v} \times (\mathbf{E} + \mathbf{v} \times \mathbf{B}). \tag{10.12}$$

Consider now, in an aether frame, an electromagnetic shock propagating through the dielectric in the absence of any *free* charges or currents; the *bound* charges and currents are of course a matter to be decided by the response functions (10.11). As in §7, it is easy to check that the shock must be moving; otherwise all jumps vanish. Again, it is sufficient to consider the two jump conditions

$$\mathbf{n} \times [\![\mathbf{H}]\!] + u_n [\![\mathbf{D}]\!] = 0,$$

$$\mathbf{n} \times [\![\mathbf{E}]\!] - u_n [\![\mathbf{B}]\!] = 0, \tag{10.13}$$

where we have denoted the normal shock velocity by u_n in order to avoid confusing it with the velocity of the dielectric. For simplicity, we take the latter to be normal to the shock. If we now substitute (10.12) in

---

† The term refers to the two kinds of charge – positive and negative – that, in an isolated, polarized body must cancel each other exactly, as we have seen. In the old literature a dielectric is defined as an insulator, i.e. a material that will not allow the passage of *free* electric current. In such a material, only a *displacement current* $\mathbf{P}_t$ or a *magnetic current* $\mathbf{curl\ M}$ are possible $((9.2)_2)$. But engineers commonly speak of 'lossy dielectrics', which would strictly mean 'conducting insulators'. We shall use the term dielectric to describe any material capable of polarization.

$(10.13)_1$ we arrive, after a simple calculation, at the following condition for non-zero jumps:

$$u_n^2 + \chi(u_n - v)^2 - c^2 = 0. \tag{10.14}$$

This is a quadratic equation for u_n. For a dielectric at rest (in the aether frame) it gives for the velocity of the shock

$$u_n = \pm\frac{c}{n}, \qquad n = \sqrt{1 + \chi}. \tag{10.15}$$

The two possible velocities are smaller than the speed of light c in vacuum by the factor n, called the *index of refraction* of the dielectric. If the dielectric is moving, the solution (to the first order in v/c) is

$$u_n = \pm\frac{c}{n} + (1 - \frac{1}{n^2})v. \tag{10.16}$$

This dragging of light by a moving dielectric was predicted by Fresnel and confirmed by Fizeau in an experiment that used flowing water. Equation (10.16) shows that the dragging is imperfect, because only a fraction of v is added to $\pm c/n$.

CHAPTER IV

The Fusion of Electromagnetism

with Mechanics

and Thermodynamics

In the previous chapters we have laid down three principles which have led us to various relations between the electromagnetic fields, culminating in Maxwell's equations for polarizable and magnetizable materials. But these equations do not tell us what the electromagnetic fields are supposed to do. This lack of any connection between the laws of electromagnetism and the rest of physics has also forced us to introduce separate dimensions, with undetermined units, for charge and magnetic flux. Since electric motors, radios and electric kettles are a part of our everyday lives, it will not come as a great surprise if we say that electromagnetic fields have dynamical and thermal effects. But the precise way in which electromagnetism is linked to mechanics and thermodynamics is far from trivial. It must be done with great care and therefore requires some patience. We begin with a review of continuum mechanics. In this classical theory all tensors will be three-dimensional Cartesian tensors. These are tensors with respect to orthogonal space transformations only, and for them the difference between covariance and contravariance disappears. Hence subscripts suffice.

11. The laws of continuum mechanics

In the mechanics of mass points the trajectory of the ith mass point is given by specifying its position as a function of the time, viz. $\mathbf{x}_i(t)$.

In a continuous body each material point has a trajectory, too, but the material points cannot be labelled by an integer. A method of labelling the material points of a continuous body was devised by Euler (see Figure 3). Let $\mathbf{X} = (X_1, X_2, X_3)$ be the position of a material point at $t = 0$. Then the triad (X_1, X_2, X_3) can serve as a label for this material point, and its trajectory given by $\mathbf{x}(\mathbf{X}, t)$. It is not difficult to suggest other methods of labelling: there is nothing to prevent us from choosing a time different from $t = 0$; we can use any other triad (Y_1, Y_2, Y_3) related to (X_1, X_2, X_3) by a one-to-one mapping; and so on. But for our purposes, Euler's method will do. Given the vector function $\mathbf{x}(\mathbf{X}, t)$, we obtain the trajectory of a material point $\mathbf{X}$ by keeping $\mathbf{X}$ fixed and letting t vary. Its velocity and acceleration will be given by the partial time derivatives

$$\dot{\mathbf{x}} = \frac{\partial \mathbf{x}(\mathbf{X}, t)}{\partial t}, \qquad \ddot{\mathbf{x}} = \frac{\partial^2 \mathbf{x}(\mathbf{X}, t)}{\partial t^2}. \tag{11.1}$$

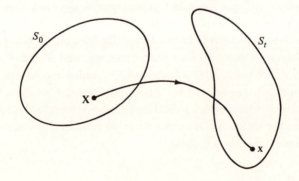

FIGURE 3

We can also fix t and vary $\mathbf{X}$ in $\mathbf{x}(\mathbf{X}, t)$. This gives us the positions of the various material points $\mathbf{X}$ at time t – a *picture* of the body at time t. The nine partial derivatives,

$$F_{ij}(\mathbf{X}, t) = \frac{\partial x_i(\mathbf{X}, t)}{\partial X_j}, \tag{11.2}$$

form the *deformation tensor*, or *matrix*. Its determinant is the Jacobian of the transformation from $\mathbf{X}$ to $\mathbf{x}$, and we shall assume that it never vanishes. This ensures the invertibility of $\mathbf{x}(\mathbf{X}, t)$: given $\mathbf{x}$ in the body, there exists one material point $\mathbf{X}$ which is there at time t. In Euler's method $\det F = 1$ at $t = 0$.

The mass m of a body is assumed to be a continuous function of its volume (cf. (1.1)):

$$m = \int \rho \, dV. \tag{11.3}$$

Of course, the density ρ is positive.†

The first principle of continuum mechanics is the conservation of mass. Thus the integral in (11.3) is conserved, but since the body may be moving and deforming, the limits of the integral may be changing, and it does not follow that ρ is constant. We can, however, change to an integral over the X_i by writing $\int \rho \, d^4x = \int \rho \det F \, d^4X$. Now the limits are constant, and the law of mass conservation, in local form, is

$$\rho(\mathbf{X}, t) \det F(\mathbf{X}, t) = \rho_0(\mathbf{X}). \tag{11.4}$$

This is Euler's equation of continuity.

PROBLEM

Problem 11–1. By differentiating (11.4) and using (3.3), prove that

$$\dot{\rho} + \rho \operatorname{tr} \dot{F} F^{-1} = 0. \tag{11.5}$$

† As in the case of electric charge, the assumption (11.3) is easily relaxed. It is only for reasons of economy that we do not mention distributions of mass along lines, surfaces, etc.

Besides the *material description* of any quantity f as a function $A(\mathbf{X}, t)$ of $\mathbf{X}$ and t, there is the *spatial description*, in which the same quantity is regarded as a function $a(\mathbf{x}, t)$ of position $\mathbf{x}$ and t.† The connection between them is, of course, that at time t the position $\mathbf{x}$ is one occupied by a material point $\mathbf{X}$ such that $\mathbf{x} = \mathbf{x}(\mathbf{X}, t)$. The material time derivative

$$\dot{f} = \frac{\partial A(\mathbf{X}, t)}{\partial t} \tag{11.6}$$

is connected with the partial derivatives of the spatial function $a(\mathbf{x}, t)$: if we use the chain rule in $A(\mathbf{X}, t) = a(\mathbf{x}(\mathbf{X}, t), t)$ we obtain $\dot{f} = a_t + \dot{\mathbf{x}} \cdot \mathbf{grad}\, a$. In Euler's notation,

$$\dot{f} = f_t + \dot{\mathbf{x}} \cdot \mathbf{grad}\, f. \tag{11.7}$$

Of particular interest are the nine space-derivatives of the velocity in the spatial description:

$$(\mathrm{grad}\, \dot{\mathbf{x}})_{ij} = \frac{\partial \dot{x}_i(\mathbf{x}, t)}{\partial x_j}; \tag{11.8}$$

this tensor, or matrix, is called the *velocity gradient*. Its trace is $\mathrm{div}\, \dot{\mathbf{x}}$.

PROBLEMS

Problem 11–2. Prove that

$$\ddot{\mathbf{x}} = \dot{\mathbf{x}}_t + (\mathrm{grad}\, \dot{\mathbf{x}})\dot{\mathbf{x}}, \tag{11.9}$$

where the last term is the vector obtained by multiplying the matrix $\mathrm{grad}\, \dot{\mathbf{x}}$ by the vector $\dot{\mathbf{x}}$.

Problem 11–3. Prove that

$$\mathrm{grad}\, \dot{\mathbf{x}} = \dot{F}F^{-1}. \tag{11.10}$$

From (11.5) and (11.10) we obtain

$$\dot{\rho} + \rho\, \mathrm{div}\, \dot{\mathbf{x}} = 0, \tag{11.11}$$

† The material description, introduced by Euler, is usually called 'Lagrangian'; the spatial description, due to d'Alambert, is usually called 'Eulerian'.

another Euler's equation of continuity expressing mass conservation. Equation (11.4) is of course a first integral of (11.11).

PROBLEM

Problem 11–4. Deduce the equation of continuity in the spatial description:

$$\rho_t + \operatorname{div} \rho \dot{\mathbf{x}} = 0. \tag{11.12}$$

We have expressed the global mass conservation law in various local forms. These are of course only valid where ρ is smooth. On a surface across which ρ is discontinuous, mass conservation requires the jump condition

$$[\![\rho(\dot{\mathbf{x}} - \mathbf{v})]\!] \cdot \mathbf{n} = 0, \tag{11.13}$$

where $\mathbf{v}$ is the velocity of the surface.

The laws of mechanics require for their formulation the concepts of momentum, force, angular momentum (or moment of momentum) and torque. They are defined, respectively, by

$$\mathbf{G} = \int \dot{\mathbf{x}}\, dm,$$

$$\mathbf{F} = \oint \mathbf{t}\, dS + \int \mathbf{b}\, dm,$$

$$\mathbf{L}_O = \int (\mathbf{x} - \mathbf{x}_O) \times \dot{\mathbf{x}}\, dm,$$

$$\mathbf{M}_O = \oint (\mathbf{x} - \mathbf{x}_O) \times \mathbf{t}\, dS + \int (\mathbf{x} - \mathbf{x}_O) \times \mathbf{b}\, dm. \tag{11.14}$$

Note that the force is assumed to consist of two kinds: a *surface*, or *contact*, force $\mathbf{t}$ per unit area of the surface of the body; and a *body* force $\mathbf{b}$ per unit mass, which is the same as $\rho\mathbf{b}$ per unit volume. As an example of a surface force, we cite the inward pointing pressure force $-p\mathbf{n}\, dS$ on a surface element dS, where p (the pressure) is the force per unit area. Of course $\mathbf{t}\, dS$ need not in general be normal to the surface. A common example of a body force is provided by gravity; then $\mathbf{b}$ is simply the acceleration of gravity. The angular momentum and torque are defined with respect to a point O with position $\mathbf{x}_O$.

We now assume that there exists a frame, called inertial, in which for a fixed point O

$$\dot{\mathbf{G}} = \mathbf{F}, \qquad \dot{\mathbf{L}}_O = \mathbf{M}_O. \tag{11.15}$$

These are Euler's laws of mechanics.†

PROBLEMS

Problem 11–5. Use $(11.15)_1$ to show that the forces exerted by any two bodies on each other are equal and opposite (whether or not there are any other forces acting on these bodies).

Problem 11–6. Use both of Euler's laws to show that the line of action of the force that a *mass point* exerts on *another* is the line joining them.

If we substitute (11.14) in Euler's laws, we obtain

$$\frac{d}{dt} \int \dot{\mathbf{x}}\,dm = \oint \mathbf{t}\,dS + \int \mathbf{b}\,dm,$$

$$\frac{d}{dt} \int (\mathbf{x} - \mathbf{x}_O) \times \dot{\mathbf{x}}\,dm =$$

$$\oint (\mathbf{x} - \mathbf{x}_O) \times \mathbf{t}\,dS + \int (\mathbf{x} - \mathbf{x}_O) \times \mathbf{b}\,dm. \tag{11.16}$$

From $(11.16)_1$ it follows from Cauchy's theorem that, whenever $\mathbf{t}$ is smooth, its components are linear and homogeneous functions of the normal $\mathbf{n}$ to the surface (cf. (1.3)). Hence there exist nine functions T_{ij} – the components of the *stress tensor* – such that, in matrix notation,

$$\mathbf{t} = T\mathbf{n}. \tag{11.17}$$

† Announced by him in the seventeen seventies, towards the end of his life. The law $m\ddot{\mathbf{x}} = \mathbf{F}$ was also announced by Euler in 1752, and was hailed as a great discovery, but he later realized that it was insufficient. In those days Newton's writings were still read, and it occurred to no one that Euler's $m\ddot{\mathbf{x}} = \mathbf{F}$ was contained in, or anticipated by, Newton's Second Law. In fact, this formula is nowhere to be found in Newton's writings.

As in §5, we shall assume all quantities to be piecewise smooth. It then follows that, wherever the quantities are smooth, the integral law $(11.16)_1$ is equivalent to Cauchy's first equation,

$$\rho \ddot{\mathbf{x}} = \mathbf{div}\, T + \rho \mathbf{b}, \tag{11.18}$$

where $\mathbf{div}\, T$ denotes the vector whose ith component is $(\mathbf{div}\, T)_i = \partial T_{ij}/\partial x_j$. Across a surface of discontinuity whose velocity is $\mathbf{v}$, the integral law leads to the jump conditions

$$[\![\rho \dot{\mathbf{x}} \otimes (\dot{\mathbf{x}} - \mathbf{v}) - T]\!]\mathbf{n} = 0. \tag{11.19}$$

Equation (11.19) is also written in matrix notation and has three components. In addition, we have used the notation for the direct, or outer, product:

$$A = \mathbf{b} \otimes \mathbf{c} \qquad \text{means} \qquad A_{ij} = b_i c_j. \tag{11.20}$$

Euler's second law $(11.16)_2$ can now be shown to lead to Cauchy's second equation,

$$T^T = T, \tag{11.21}$$

where T^T is the transpose of T. Thus T is symmetric.

We now turn to the principles of thermodynamics. These require some additional primitives. The first is the *energy* E, which we assume, as usual, to be a continuous function of the mass, or of the volume, of a body:†

$$E = \int \varepsilon \, dm = \int \rho \varepsilon \, dV. \tag{11.22}$$

We warn against confusing this primitive with a first integral of the laws of mechanics. Such a first integral, even if called energy, would be a derived quantity, not a primitive. (We trust that the energy E will not easily be confused with the electric field $\mathbf{E}$.)

Another primitive is *heating* $\mathcal{Q}$ of a body, which we shall presently assume to contribute to the rate of change of its energy. We refrain from

---

† This assumption may need to be relaxed; cf. the first footnote of this section. Even for a volume-filling body, a surface integral should be added to (11.22) if surface tension is important.

saying that $\mathcal{Q}$ is the rate at which a body receives heat, because $\mathcal{Q}$ is not assumed to be the time derivative of anything. In continuum mechanics heating is generally taken to be a combination of *surface*, or *contact*, heating β per unit area and *body* heating h per unit mass (or ρh per unit volume):

$$\mathcal{Q} = \oint \beta \, dS + \int h \, dm. \tag{11.23}$$

In interpreting the theory, we might think of the first term as describing conduction of heat through the boundary of the body, and of the second as representing a distribution of heat sources inside the body, perhaps through the absorption of some kind of radiation. Heating may be negative, in which case it is called *cooling*. It may also vanish, and if it does so over a time interval, the body is said to undergo an *adiabatic* process.

The third concept we need – not a primitive – is the power of the forces acting on the body:

$$\Pi = \oint \dot{\mathbf{x}} \cdot T\mathbf{n} \, dS + \int \dot{\mathbf{x}} \cdot \mathbf{b} \, dm. \tag{11.24}$$

The first principle of thermodynamics states that

$$\dot{E} = \Pi + \mathcal{Q}. \tag{11.25}$$

If we substitute (11.22)–(11.24), this becomes

$$\frac{d}{dt} \int \varepsilon \, dm = \oint \dot{\mathbf{x}} \cdot T\mathbf{n} \, dS + \int \dot{\mathbf{x}} \cdot \mathbf{b} \, dm + \oint \beta \, dS + \int h \, dm. \tag{11.26}$$

By Cauchy's theorem, β must be of the form $\beta = -\mathbf{q} \cdot \mathbf{n} = -q_n$, where $\mathbf{q}$ is called the *heat flux*; the sign is conventional. (The use of the same letter for the heat flux vector $\mathbf{q}$ and for the electric charge density q is unfortunate, but we regard the danger of confusing the two as unlikely.) Thus

$$\frac{d}{dt} \int \varepsilon \, dm = \oint \dot{\mathbf{x}} \cdot T\mathbf{n} \, dS + \int \dot{\mathbf{x}} \cdot \mathbf{b} \, dm - \oint q_n \, dS + \int h \, dm. \tag{11.27}$$

This statement of the first principle, or axiom, of thermodynamics is also (appropriately) called *the law of energy balance*. It leads, as usual, to the local law

$$\rho \dot{\varepsilon} = \operatorname{div} \dot{\mathbf{x}} T + \rho \dot{\mathbf{x}} \cdot \mathbf{b} - \operatorname{div} \mathbf{q} + \rho h, \tag{11.28}$$

where $(\dot{\mathbf{x}}T)_j = \dot{x}_i T_{ij}$, and to the jump condition

$$\llbracket \rho\varepsilon(\dot{\mathbf{x}} - \mathbf{v}) - \dot{\mathbf{x}}T + \mathbf{q} \rrbracket \cdot \mathbf{n} = 0. \tag{11.29}$$

We shall assume that the heat flux vector $\mathbf{q}$ is a Galilei invariant. Formally, this can be achieved by defining its four-dimensional components to be $(\mathbf{q}, 0)$ in any one inertial frame. Such a (four-dimensional) vector is called a *space vector*.

Two further primitives are needed for the formulation of the second principle of thermodynamics. The first is *temperature* ϑ, a property we associate with the material points of a body. Like density, it has a material description $\vartheta = A(\mathbf{X}, t)$ or a spatial description $\vartheta = a(\mathbf{x}, t)$, the latter being equal to the former for the material point $\mathbf{X}$ passing through $\mathbf{x}$ at time t. It is meant to describe the hotness (or coldness) of material points. We shall further assume that temperature has a universal lower bound below which no body can ever be cooled. Of course, the assumption of a universal lower bound is really an axiom. It is analogous to the axiom, which we seldom bother to mention, that mass is always positive, in all frames and in every system of units. If we measure ϑ from this lower bound, as we shall, it becomes positive.†

The second primitive is the *entropy* S of a body. We assume that, like energy, it is a continuous function of mass:‡

$$S = \int \eta \, dm, \tag{11.30}$$

where η is the entropy per unit mass, or the *specific entropy*.

The second principle of thermodynamics states that the entropy increases at least as fast as the sum of heatings, each divided by the temperature at which it takes place:

$$\dot{S} = \frac{d}{dt} \int \eta \, dm \geq \int \frac{h}{\vartheta} \, dm - \oint \frac{q_n}{\vartheta} \, dS. \tag{11.31}$$

---

† In the kinetic theory of gases the foregoing analogy becomes a theorem: temperature is there *defined* to be the average kinetic energy of a molecule; it is then positive *because* mass is positive.

‡ Again, this may need to be relaxed. Cf. the first and fourth footnotes of this section.

Of course, $\dot{S}$ need not be positive. If the inequality in (11.31) becomes an equality throughout a time interval, we say that the body is undergoing a *reversible process* during that time interval. A process that is not reversible is *irreversible*.

For piecewise smooth functions the integral inequality (11.31) reduces to the local inequality

$$\rho\dot{\eta} \geq h/\vartheta - \operatorname{div}(\mathbf{q}/\vartheta), \tag{11.32}$$

and to the jump inequality

$$[\![\rho\eta(\dot{\mathbf{x}} - \mathbf{v}) + \mathbf{q}/\vartheta]\!] \cdot \mathbf{n} \geq 0, \tag{11.33}$$

which is – like all the other jump conditions – subject to the prescription following (5.9).

While the law of energy balance, like the laws of mechanics, is assumed to hold everywhere and at all times in all bodies, the second principle of thermodynamics does not enjoy the same status. Bodies may exist that have no unique temperature, or even no temperature at all. They lie outside the scope of thermodynamics. From now on, when we speak of ϑ, the qualifying phrase 'if it exists' will be understood.

When we say that a body has a temperature ϑ, rather than a temperature field $\vartheta(\mathbf{x}, t)$, we imply that its temperature is uniform, and is at most a function of the time. For such bodies, which almost exhaust the subject matter of treatises on classical thermodynamics, the temperature can be taken outside the integrals of (11.31), and the entropy inequality becomes simply $\dot{S} \geq \mathcal{Q}/\vartheta$.

12. Continuum mechanics and electromagnetism

In the previous section we have formulated five principles of continuum mechanics: conservation of mass, Euler's two laws, the law of energy balance and the entropy inequality. A theory capable of describing the dynamic and thermal effects of electromagnetic fields requires a certain generalization of these principles. Finally, of course, the principles of electromagnetism must be added.

We generalize, or modify, the laws of continuum mechanics in two places. First, we employ a more general expression than the velocity $\dot{\mathbf{x}}$

for the momentum per unit mass, and denote it by $\mathbf{g}$. Thus we take $\mathbf{G}$ and $\mathbf{L}_O$ to be

$$\mathbf{G} = \int \mathbf{g}\,dm, \qquad \mathbf{L}_O = \int (\mathbf{x} - \mathbf{x}_O) \times \mathbf{g}\,dm. \qquad (12.1)$$

With this modification, Euler's first law, $\dot{\mathbf{G}} = \mathbf{F}$, leads to the equation

$$\rho\dot{\mathbf{g}} = \mathbf{div}\,T + \rho\mathbf{b} \qquad (12.2)$$

and to the three jump conditions

$$[\![\rho\mathbf{g} \otimes (\dot{\mathbf{x}} - \mathbf{v}) - T]\!]\mathbf{n} = 0. \qquad (12.3)$$

His second law, $\dot{\mathbf{L}}_O = \mathbf{M}_O$, leads to the three equations

$$\rho\dot{x}_{[i}g_{j]} + T_{[ij]} = 0, \qquad (12.4)$$

where the square brackets mean that the expression with ji is to be subtracted from the expression with ij (cf. Problem 2–4). Note that, unless $\dot{x}_i g_j = \dot{x}_j g_i$, the stress tensor is no longer symmetric. Equations (12.2)–(12.4) replace (11.18), (11.19) and (11.21).

We shall keep an open mind as to what $\mathbf{g}$ is, but if the modified theory is to be a generalization of the original one, we must clearly impose the condition that $\mathbf{g}$ reduce to $\dot{\mathbf{x}}$ in the absence of an electromagnetic field.

The second place where we make a change is in the balance of energy: we introduce, alongside the heat flux vector $\mathbf{q}$, an extra energy flux vector equal to $\mathcal{E} \times \mathcal{H}$, the vector product of the electromotive intensity $\mathcal{E} = \mathbf{E} + \dot{\mathbf{x}} \times \mathbf{B}$ and the magnetomotive intensity $\mathcal{H} = \mathbf{B}/\mu_0 - \dot{\mathbf{x}} \times \epsilon_0\mathbf{E} - \mathcal{M}$†
With this extra term, (11.28)–(11.29) are replaced by

$$\rho\dot{\varepsilon} = \mathrm{div}\,\dot{\mathbf{x}}T + \rho\dot{\mathbf{x}} \cdot \mathbf{b} - \mathrm{div}(\mathbf{q} + \mathcal{E} \times \mathcal{H}) + \rho h, \qquad (12.5)$$

$$[\![\rho\varepsilon(\dot{\mathbf{x}} - \mathbf{v}) - \dot{\mathbf{x}}T + \mathbf{q} + \mathcal{E} \times \mathcal{H}]\!] \cdot \mathbf{n} = 0. \qquad (12.6)$$

---

† This assumption is motivated by applications of Poynting's theorem (which we shall derive below; cf. (14.6)) to special situations, and on identifications of various expressions as energy, mechanical power or heating. From such considerations it appears plausible that $\mathcal{E} \times \mathcal{H}$, plus the curl of an unspecified vector, represents an energy flux. A final assumption is then made to the effect that $\mathcal{E} \times \mathcal{H}$ is itself an energy flux. We prefer an unambiguous and frank statement of this hypothesis.

Two features of this modification of the law of energy balance should be noted. First, it is obviously a generalization, because $\mathcal{E} = \mathbf{E} + \dot{\mathbf{x}} \times \mathbf{B}$ vanishes whenever both $\mathbf{E}$ and $\mathbf{B}$ do. Secondly, by asserting that $\mathcal{E} \times \mathcal{H}$ is an energy flux (energy per unit area per unit time), we have made a connection between the dimensions of electric charge and magnetic flux and the mechanical dimensions. The product $\mathcal{E} \times \mathcal{H}$ has the dimensions $\Phi C/(LT)^2$; we now find that these are also the dimensions of velocity times force per unit area. Hence ΦC is dimensionally the same as NLT, where N is the dimension of force. Consider now the constant μ_0, with dimensions $(\Phi/C)/(L/T)$. According to the relation we have just established, the dimensions of μ_0 are $N/(C/T)^2$, i.e. force divided by squared current. But μ_0 is a *universal constant*. Expressed in terms of *known* units of force and current, it will have a definite numerical value, to be determined by a suitable experiment. Conversely, an *assumed* value for μ_0 will serve to fix the unit of current in terms of the unit of force (just as the metre was fixed by (7.9) in terms of the second). This is the method by which the unit of current, the *ampere*, is fixed in SI units. Specifically, we set

$$\mu_0 = 4\pi \times 10^{-7} \text{ newton/amp}^2 \tag{12.7}$$

(exactly) as the SI definition of the ampere. If this sounds too theoretical, we need only remark that the above mentioned 'suitable experiment' will serve to measure currents in amperes. We shall return to this point after we are in possession of a formula for the force on a current-carrying body.

All the electromagnetic SI units follow from this unit of current. Thus the unit of charge, the *coulomb*, is defined by 1 coulomb = 1 amp · s; the unit of electric potential, the *volt*, by 1 volt = 1 newton · m/coulomb; the unit of $\mathbf{B}$, the *tesla*, by 1 tesla = 1 newton/(amp · m); the unit of magnetic flux, the *weber*, by 1 weber = 1 tesla · m^2; etc.

We can also use (12.7) to determine the numerical value of the second universal constant, ϵ_0, in these units:

$$\frac{1}{4\pi\epsilon_0} = \frac{\mu_0 c^2}{4\pi} = 10^{-7} (\text{ newton/amp}^2) \cdot (3 \times 10^8 \text{ m/s})^2$$

$$= 9 \times 10^9 \text{ newton} \cdot \text{m}^2/\text{coulomb}^2. \tag{12.8}$$

Of course, the factor 9 (not the *exponent* 9) is 3^2; a more accurate value of c would lead to a more accurate value of this factor.

It is important to realize that the modifications we have made in the laws of mechanics and in the law of energy balance will force us to give up some of our pre-electromagnetic notions regarding motion, forces and energy. While, according to $\dot{\mathbf{G}} = \mathbf{F}$, imbalance of applied forces is still the same as change of momentum, the latter is *not* the same as acceleration (even in a rigid body), for $\mathbf{G}$ is no longer $\int \dot{\mathbf{x}}\, dm$. Similarly, $\dot{\mathbf{L}}_O = \mathbf{M}_O$; hence imbalance of torques is still the same as change of angular momentum, but the latter is no longer directly related to change in angular velocity (even in a rigid body). Finally, change of energy is no longer the same as imbalance between mechanical power and heating, because there is now the extra term $\int -\boldsymbol{\mathcal{E}} \times \boldsymbol{\mathcal{H}} \cdot \mathbf{n}\, dS$.

We make no change in the law of mass conservation or in the entropy inequality.

It is time to make a list of all the classical laws that govern the behaviour of continuous bodies. For reasons which will become clear later, we list only the local equations, leaving out the jump conditions for the present:

$$\dot{\rho} + \rho \operatorname{div} \dot{\mathbf{x}} = 0, \tag{12.9}$$

$$\rho\dot{\mathbf{g}} = \operatorname{div} T + \rho\mathbf{b}, \tag{12.10}$$

$$\rho\dot{x}_{[i}g_{j]} + T_{[ij]} = 0, \tag{12.11}$$

$$\rho\dot{\varepsilon} = \operatorname{div} \dot{\mathbf{x}}T + \rho\dot{\mathbf{x}} \cdot \mathbf{b} + \rho h - \operatorname{div}(\mathbf{q} + \boldsymbol{\mathcal{E}} \times \boldsymbol{\mathcal{H}}), \tag{12.12}$$

$$\rho\dot{\eta} \geq \rho h/\vartheta - \operatorname{div}(\mathbf{q}/\vartheta), \tag{12.13}$$

$$\operatorname{div} \mathbf{D} = q, \tag{12.14}$$

$$\operatorname{\mathbf{curl}} \boldsymbol{\mathcal{H}} = \boldsymbol{\mathcal{J}} + \overset{*}{\mathbf{D}}, \tag{12.15}$$

$$\operatorname{div} \mathbf{B} = 0, \tag{12.16}$$

$$\operatorname{\mathbf{curl}} \boldsymbol{\mathcal{E}} = -\overset{*}{\mathbf{B}}, \tag{12.17}$$

$$\mathbf{D} = \epsilon_0\mathbf{E} + \mathbf{P}, \tag{12.18}$$

$$\boldsymbol{\mathcal{H}} = \mathbf{B}/\mu_0 - \dot{\mathbf{x}} \times \epsilon_0\mathbf{E} - \boldsymbol{\mathcal{M}}. \tag{12.19}$$

These laws, it will be recalled, hold in an inertial frame that is also an aether frame. They are obviously under-determined and must be supplemented by further information. This information takes the form of *constitutive relations* for the various material properties. We have, in fact, already met an example of a constitutive relation in the response function (10.11) of a classical, linear dielectric. We shall now have to consider constitutive relations for various other quantities that we regard as determined by material properties, such as the three components of the specific momentum $\mathbf{g}$, the nine components of the stress T, the specific energy ε, and so on. Not every quantity appearing in (12.9)–(12.19) is of this kind. Ocean tides, for example, are the result of an imbalance between $\mathbf{div}\ T$ and a body force $\rho\mathbf{b}$ arising from the gravitational pull of the earth, the sun and the moon. But, whereas the stress T is a material property – depending, presumably, on the material constitution and on its state (density, temperature, etc.) – the proximity of the moon is not a property of sea water. Thus $\mathbf{b}$ is *not* to be determined by a constitutive relation. In the next section we shall discuss these matters in a more general context.

13. Processes and constitutive relations

The laws (12.9)–(12.13) are written in terms of material derivatives. This is of course a matter of choice, since we can always use Euler's formula (11.7) to express a material derivative in terms of partial time derivatives. The same choice is available in Maxwell's equations, because the flux derivative $\overset{*}{\mathbf{A}}$ that we defined in (5.4) is related to the material derivative $\dot{\mathbf{A}}$ through

$$\overset{*}{\mathbf{A}} = \dot{\mathbf{A}} + \mathbf{A}\operatorname{div}\dot{\mathbf{x}} - (\mathbf{A}\cdot\mathbf{grad})\dot{\mathbf{x}}, \qquad (13.1)$$

again by Euler's formula (11.7). In what follows we shall find it convenient to use the material description.

We define a *process* as a collection of the ten functions $\mathbf{x}(\mathbf{X}, t)$, $\vartheta(\mathbf{X}, t)$, $\mathbf{E}(\mathbf{X}, t)$ and $\mathbf{B}(\mathbf{X}, t)$, the latter subject to $\operatorname{div}\mathbf{B} = 0$. Since these are not all the quantities that enter the various laws, we really ought to use a qualifying word, and talk about (say) an electromagnetothermokinetic process, but this is awkward. It is simpler to talk about an $(\mathbf{x}, \vartheta, \mathbf{E}, \mathbf{B})$-process, or just a process.

We shall now assume a constitutive relation of the form

$$f = \mathcal{F}[\mathbf{x}, \vartheta, \mathbf{E}, \mathbf{B}] \tag{13.2}$$

for each of the twenty-six quantities

$$\mathbf{g}, \quad T, \quad \varepsilon, \quad \mathbf{q}, \quad \eta, \quad \mathcal{J}, \quad \mathbf{P} \quad \text{and} \quad \mathcal{M} \tag{13.3}$$

The square brackets in (13.2) mean that the right hand side is a *functional* rather than a function. It may depend not just on the values of its arguments at $\mathbf{X}$ and t, but also on their *histories* up to t, not only at $\mathbf{X}$, but also at other material points. For example, the $\mathcal{F}$'s may depend on components of the temperature gradient. We shall assume that each of the twenty-six $\mathcal{F}$'s is differentiable, as often as is needed.

We note that, in accordance with the remarks made at the end of the preceding section, the specific body force $\mathbf{b}$ does not appear in the list of quantities (13.3). Similarly, the heat flux vector $\mathbf{q}$, which we consider to be a material property, appears in (13.3), but the specific body heating h does not. We shall presently see that these omissions lead to a meaningful combination of processes, constitutive relations and physical laws.

Consider now an $(\mathbf{x}, \vartheta, \mathbf{E}, \mathbf{B})$-process, together with the collection of twenty-six assumed constitutive relations. They will both be called *admissible* if they satisfy all the laws, *except* for Euler's second law (12.11) and the entropy inequality (12.13). In spite of the name, admissibility need not be a severe restriction on the process and the constitutive relations. Thus, the fulfillment of the law of mass conservation (12.9) just means that $\dot{\rho}$ is determined by the process, because $\mathbf{x}(\mathbf{X}, t)$ determines $\operatorname{div} \dot{\mathbf{x}} = \operatorname{tr} F^{-1}\dot{F}$. With $\mathbf{g}$ and T given by differentiable constitutive relations, Euler's first law (12.10) becomes a formula for the body force $\mathbf{b}$. Similarly, with $\mathbf{b}$ fixed in this way, the law of energy balance (12.12) becomes a formula for the body heating h. Equation (12.14) is a formula for the charge density q, since $\mathbf{D}$ is fixed by the process and the constitutive relations in accordance with the aether relation (12.18). Equation (12.16) is part of the definition of a process. Equations (12.15) and (12.19) serve to determine $\dot{\mathbf{E}}$, and (12.17) determines $\dot{\mathbf{B}}$.

The fact that an admissible process, together with a given set of constitutive relations, only requires suitable body forces **b** and body heatings h reflects our experience. After all, a restriction on $\mathbf{x}(\mathbf{X}, t)$ or $\vartheta(\mathbf{X}, t)$, say, would mean that a body exists which cannot be taken from A, where its temperature is given, and brought to B with another assigned temperature. Of course (we say from experience), this may require some pushing or pulling (suitable **b**) and heating or cooling (suitable h), but it can be done.

What about the laws (12.11) and (12.13), that we have left out? Clearly, since h has already been fixed and all other quantities that appear in these laws are determined, either by the process or by the constitutive relations, they impose real restrictions, either on the processes or on the constitutive relations. Faced with this mathematical conclusion, we make a physical choice, based on experience: the processes are not to be restricted. We are then forced, by the entropy inequality and by Euler's second law of mechanics, to accept restrictions on the constitutive relations. Speaking somewhat loosely, we say: all processes are possible, but not every material imaginable exists.

The foregoing method, in a rather more special setting, was suggested by Coleman and Noll in 1963. In the next chapter we shall illustrate it by a few examples.

14. Reduction of the entropy inequality

In order to implement the method of Coleman and Noll, we shall first eliminate the body force **b** and the body heating h in accordance with the foregoing discussion. We begin by substituting **b** from (12.10) into (12.12). This gives

$$\rho\dot{\varepsilon} = \operatorname{div}\dot{\mathbf{x}}T + \dot{\mathbf{x}} \cdot (\rho\dot{\mathbf{g}} - \operatorname{\mathbf{div}} T) + \rho h - \operatorname{div}(\mathbf{q} + \boldsymbol{\mathcal{E}} \times \boldsymbol{\mathcal{H}})$$

$$= T_{ij}(\operatorname{grad}\dot{\mathbf{x}})_{ij} + \rho(\dot{\mathbf{g}} \cdot \dot{\mathbf{x}} + h) - \operatorname{div}(\mathbf{q} + \boldsymbol{\mathcal{E}} \times \boldsymbol{\mathcal{H}}). \qquad (14.1)$$

It is convenient to define, for any two matrices A and B, a *scalar product* by

$$A \cdot B = \operatorname{tr} A^T B = A_{ij}B_{ij}. \qquad (14.2)$$

It can be shown that the name is justified, because this product has all the properties of a scalar product. For example, $B \cdot A = A \cdot B$; if $A \cdot A = 0$ then $A = 0$; if $A \cdot X = 0$ for all X, then $A = 0$. With this notation, we can write (14.1) in the form

$$\rho \dot{\varepsilon} = T \cdot \operatorname{grad} \dot{\mathbf{x}} + \rho(\dot{\mathbf{g}} \cdot \dot{\mathbf{x}} + h) - \operatorname{div}(\mathbf{q} + \boldsymbol{\mathcal{E}} \times \boldsymbol{\mathcal{H}}). \tag{14.3}$$

Next, we substitute h from this equation into the entropy inequality (12.13):

$$\rho \dot{\eta} \geq [\rho \dot{\varepsilon} - T \cdot \operatorname{grad} \dot{\mathbf{x}} - \rho \dot{\mathbf{g}} \cdot \dot{\mathbf{x}} + \operatorname{div}(\mathbf{q} + \boldsymbol{\mathcal{E}} \times \boldsymbol{\mathcal{H}})]/\vartheta - \operatorname{div}(\mathbf{q}/\vartheta). \tag{14.4}$$

We may multiply this by ϑ (since ϑ is positive) and write it as

$$\rho(\vartheta \dot{\eta} - \dot{\varepsilon} + \dot{\mathbf{g}} \cdot \dot{\mathbf{x}}) + T \cdot \operatorname{grad} \dot{\mathbf{x}} - \operatorname{div} \boldsymbol{\mathcal{E}} \times \boldsymbol{\mathcal{H}} - (\mathbf{q} \cdot \operatorname{grad} \vartheta)/\vartheta \geq 0. \tag{14.5}$$

Now, we have the identity

$$-\operatorname{div} \boldsymbol{\mathcal{E}} \times \boldsymbol{\mathcal{H}} = \boldsymbol{\mathcal{E}} \cdot \mathbf{curl}\ \boldsymbol{\mathcal{H}} - \boldsymbol{\mathcal{H}} \cdot \mathbf{curl}\ \boldsymbol{\mathcal{E}}.$$

If we substitute the curls from Maxwell's equations (12.15) and (12.17), we obtain

$$-\operatorname{div} \boldsymbol{\mathcal{E}} \times \boldsymbol{\mathcal{H}} = \boldsymbol{\mathcal{J}} \cdot \boldsymbol{\mathcal{E}} + \boldsymbol{\mathcal{E}} \cdot \overset{*}{\mathbf{D}} + \boldsymbol{\mathcal{H}} \cdot \overset{*}{\mathbf{B}}. \tag{14.6}$$

This relation (or, more especially, the one it reduces to when $\dot{\mathbf{x}} = 0$) is called Poynting's theorem. It is often, and erroneously, regarded as a balance law for 'electromagnetic energy'. Of course, it is nothing of the kind, because it is just an *identity* satisfied by all fields that are solutions of Maxwell's equations.

PROBLEMS

Problem 14–1. Prove the identity

$$\mathbf{b} \cdot \overset{*}{\mathbf{a}} = \mathbf{b} \cdot \dot{\mathbf{a}} + [(\mathbf{b} \cdot \mathbf{a})I - \mathbf{b} \otimes \mathbf{a}] \cdot \operatorname{grad} \dot{\mathbf{x}}, \tag{14.7}$$

where I is the unit matrix.

Problem 14–2. Prove the identity

$$\mathbf{a} \times \mathbf{b} \otimes \mathbf{c} + \mathbf{b} \times \mathbf{c} \otimes \mathbf{a} + \mathbf{c} \times \mathbf{a} \otimes \mathbf{b} = (\mathbf{a} \times \mathbf{b} \cdot \mathbf{c})I. \tag{14.8}$$

If we use (14.7) and (14.8), (14.6) takes the form

$$-\operatorname{div}\mathcal{E} \times \mathcal{H} = \mathcal{J} \cdot \mathcal{E} - \mathbf{P} \cdot \dot{\mathcal{E}} - \mathcal{M} \cdot \dot{\mathbf{B}} + \epsilon_0 \mathbf{E} \times \mathbf{B} \cdot \ddot{\mathbf{x}}$$

$$+[(\epsilon_0 E^2 + B^2/\mu_0 - \epsilon_0 \mathbf{E} \times \mathbf{B} \cdot \dot{\mathbf{x}} + \mathcal{E} \cdot \mathbf{P} - \mathcal{M} \cdot \mathbf{B})I$$

$$-\epsilon_0 \mathbf{E} \otimes \mathbf{E} - \mathbf{B} \otimes \mathbf{B}/\mu_0 - \mathcal{E} \otimes \mathbf{P} + \mathcal{M} \otimes \mathbf{B} - \epsilon_0 \mathbf{E} \times \mathbf{B} \otimes \dot{\mathbf{x}}] \cdot \operatorname{grad} \dot{\mathbf{x}}$$

$$+[\epsilon_0 E^2/2 + B^2/(2\mu_0) + \mathcal{E} \cdot \mathbf{P} - \epsilon_0 \mathbf{E} \times \mathbf{B} \cdot \dot{\mathbf{x}}]^{\cdot}, \tag{14.9}$$

where $[\ldots]^{\cdot}$ denotes the material derivative of $[\ldots]$. This expression is to be substituted in the inequality (14.5). Simultaneously, we define a function φ, which we call *the specific free energy*, by

$$\varphi = \varepsilon - \vartheta\eta - \mathbf{g} \cdot \dot{\mathbf{x}} + \tfrac{1}{2}\dot{x}^2$$

$$-[\epsilon_0 E^2/2 + B^2/(2\mu_0) + \mathcal{E} \cdot \mathbf{P} - \epsilon_0 \mathbf{E} \times \mathbf{B} \cdot \dot{\mathbf{x}}]/\rho. \tag{14.10}$$

In the absence of an electromagnetic field, when $\mathbf{g} = \dot{\mathbf{x}}$, φ reduces to $\varepsilon - \tfrac{1}{2}\dot{x}^2 - \vartheta\eta$. This is the familiar definition of the specific free energy in classical thermodynamics (where $\varepsilon - \tfrac{1}{2}\dot{x}^2$ is called the *internal* specific energy).

If we now use (14.10) in order to replace ε by φ, as well as (14.9), the entropy inequality (14.5) becomes

$$-\rho\dot{\varphi} - \rho\eta\dot{\vartheta} - (\rho\mathbf{g} - \rho\dot{\mathbf{x}} - \epsilon_0 \mathbf{E} \times \mathbf{B}) \cdot \ddot{\mathbf{x}} - \mathbf{P} \cdot \dot{\mathcal{E}} - \mathcal{M} \cdot \dot{\mathbf{B}}$$

$$+\tau \cdot \operatorname{grad} \dot{\mathbf{x}} + \mathcal{J} \cdot \mathcal{E} - (\mathbf{q} \cdot \operatorname{grad}\vartheta)/\vartheta \geq 0, \tag{14.11}$$

where

$$\tau = T + [\epsilon_0 E^2/2 + B^2/(2\mu_0) - \mathcal{M} \cdot \mathbf{B}]I$$

$$-\epsilon_0 \mathbf{E} \otimes \mathbf{E} - \mathbf{B} \otimes \mathbf{B}/\mu_0 - \mathcal{E} \otimes \mathbf{P} + \mathcal{M} \otimes \mathbf{B} - \epsilon_0 \mathbf{E} \times \mathbf{B} \otimes \dot{\mathbf{x}}. \tag{14.12}$$

In the next chapter we shall make extensive use of (14.11). We conclude this discussion with a corollary that follows immediately from this inequality. If ϑ, $\dot{\mathbf{x}}$, $\mathcal{E}$ and $\mathbf{B}$ are all constant, then

$$\rho\dot{\varphi} \leq \tau \cdot \operatorname{grad} \dot{\mathbf{x}} + \mathcal{J} \cdot \mathcal{E} - (\mathbf{q} \cdot \operatorname{grad}\vartheta)/\vartheta. \tag{14.13}$$

This sets an upper limit to the rate of increase of φ. With so many variables held constant, this statement has content only when φ depends on other variables, in addition (perhaps) to ϑ, $\dot{\mathbf{x}}$, $\mathcal{E}$ and $\mathbf{B}$. For example, φ may depend on the concentrations of substances undergoing a chemical reaction. In effect, then, the corollary usually refers to a *partial* rate of increase (in the sense of a partial derivative). If the right hand side of (14.13) vanishes – as it will for a uniformly moving insulator ($\mathcal{J} = \mathbf{q} = 0$), or whenever $\operatorname{grad}\dot{\mathbf{x}}$, $\mathcal{E}$ and $\mathbf{grad}\,\vartheta$ all vanish – then $\dot{\varphi} \leq 0$. Under these circumstances the free energy cannot increase; again, this usually refers to a partial time derivative. In this special form the corollary furnishes the basis for considerations of thermodynamic stability: let α be a variable on which φ depends, in addition to ϑ, $\dot{\mathbf{x}}$, $\mathcal{E}$ and $\mathbf{B}$, and let α_m be such that

$$\varphi_\alpha(\vartheta, \dot{\mathbf{x}}, \mathcal{E}, \mathbf{B}, \alpha_m) = 0, \qquad \varphi_{\alpha\alpha}(\vartheta, \dot{\mathbf{x}}, \mathcal{E}, \mathbf{B}, \alpha_m) > 0, \qquad (14.14)$$

so that φ has a minimum at $\alpha = \alpha_m$. Then α_m – which according to $(14.14)_1$ is a function of the other variables – is a *stable* value of α with respect to every change throughout which the other variables have fixed values and the right hand side of (14.13) vanishes. For φ cannot increase during any such change; therefore the value of α at the end cannot be different from α_m.

CHAPTER V

Electromagnetic

Materials

15. A simple class of fluids

Consider a class of materials for which the constitutive relations are all
of the form

$$f = \mathcal{F}(\rho, \vartheta, \dot{\mathbf{x}}, \boldsymbol{\mathcal{E}}, \mathbf{B}, \mathbf{grad}\,\vartheta), \tag{15.1}$$

where $\mathcal{F}$ is a function (a different one for each of the twenty-six quantities
(13.3)), *not* a functional, of its fourteen variables. Since $\rho\,\mathrm{div}\,F = \mathrm{const.}$,
this class of materials depends on $\mathbf{x}(\mathbf{X}, t)$ only through the determinant
of the derivatives $F_{ij} = \partial x_i/\partial X_j$ and through the material derivative $\dot{\mathbf{x}}$.
Similarly, it depends on ϑ and on $\mathbf{grad}\,\vartheta$, but not on time derivatives
of the temperature. This is obviously a special class of materials. We
shall refer to them as fluids, because (15.1) seems to correspond to some
of our prejudices regarding fluids. In a solid, we expect the constitutive
relations to depend on the distortion of the material – perhaps through
the various components of the deformation tensor F – rather than merely
on its density. The question, whether (15.1) is capable of describing all
fluids, or even any particular kind of fluid, has no meaning, because we
have not defined 'fluid' (or 'solid', for that matter). Such definitions
are quite complicated – and even controversial – and they certainly lie
outside the scope of this book.

The specific free energy has been defined by (14.10). If $\varepsilon(\ldots)$, $\eta(\ldots)$,
$\mathbf{g}(\ldots)$ and $\mathbf{P}(\ldots)$ are given by constitutive relations of the form (15.1),
φ, too, will be a function of the arguments $(\rho, \vartheta, \dot{\mathbf{x}}, \boldsymbol{\mathcal{E}}, \mathbf{B}, \mathbf{grad}\,\vartheta)$. By

the chain rule, its material derivative will be given by

$$\dot{\varphi} = \varphi_\rho \dot{\rho} + \varphi_\vartheta \dot{\vartheta} + \varphi_{\dot{\mathbf{x}}} \cdot \ddot{\mathbf{x}} + \varphi_{\boldsymbol{\mathcal{E}}} \cdot \dot{\boldsymbol{\mathcal{E}}} + \varphi_{\mathbf{B}} \cdot \dot{\mathbf{B}} + \varphi_{\mathbf{grad}\,\vartheta} \cdot (\mathbf{grad}\,\vartheta)^{\cdot}, \quad (15.2)$$

where $\varphi_{\dot{\mathbf{x}}}$ denotes the vector defined by $(\varphi_{\dot{\mathbf{x}}})_i = \partial\varphi/\partial\dot{x}_i$, and similarly for $\varphi_{\boldsymbol{\mathcal{E}}}$, $\varphi_{\mathbf{B}}$ and $\varphi_{\mathbf{grad}\,\vartheta}$. We substitute this in (14.11) and obtain

$$-\rho(\varphi_\vartheta + \eta)\dot{\vartheta} - (\rho\varphi_{\dot{\mathbf{x}}} + \rho\mathbf{g} - \rho\dot{\mathbf{x}} - \epsilon_0\mathbf{E} \times \mathbf{B}) \cdot \ddot{\mathbf{x}} - \rho\varphi_{\mathbf{grad}\,\vartheta} \cdot (\mathbf{grad}\,\vartheta)^{\cdot}$$

$$-(\rho\varphi_{\boldsymbol{\mathcal{E}}} + \mathbf{P}) \cdot \dot{\boldsymbol{\mathcal{E}}} - (\rho\varphi_{\mathbf{B}} + \boldsymbol{\mathcal{M}}) \cdot \dot{\mathbf{B}} + (\tau + \rho^2\varphi_\rho I) \cdot \mathrm{grad}\,\dot{\mathbf{x}}$$

$$+\boldsymbol{\mathcal{J}} \cdot \boldsymbol{\mathcal{E}} - (\mathbf{q} \cdot \mathbf{grad}\,\vartheta)/\vartheta \geq 0. \quad (15.3)$$

By choosing a suitable $(\mathbf{x}, \vartheta, \boldsymbol{\mathcal{E}}, \mathbf{B})$-process we can arbitrarily fix the arguments of the constitutive relations (15.1) at any material point at any time. This fixes all the quantities in (15.3), *except* for $\dot{\vartheta}$, $\ddot{\mathbf{x}}$, $(\mathbf{grad}\,\vartheta)^{\cdot}$, $\dot{\boldsymbol{\mathcal{E}}}$, $\dot{\mathbf{B}}$ and $\mathrm{grad}\,\dot{\mathbf{x}}$, which can still be varied at will by a further adaptation of the process (for example, by Faraday's law of induction, $\dot{\mathbf{B}}$ can be adjusted at any point by an appropriate choice of the space derivatives of $\boldsymbol{\mathcal{E}}$). If the inequality (15.3) is to hold for *arbitrary* processes, the factors of $\dot{\vartheta}$, $\ddot{\mathbf{x}}$, $(\mathbf{grad}\,\vartheta)^{\cdot}$, $\dot{\boldsymbol{\mathcal{E}}}$, $\dot{\mathbf{B}}$ and $\mathrm{grad}\,\dot{\mathbf{x}}$ must all vanish. Hence

$$\eta = -\varphi_\vartheta, \quad (15.4)$$

$$\mathbf{g} = \dot{\mathbf{x}} + \epsilon_0\mathbf{E} \times \mathbf{B}/\rho - \varphi_{\dot{\mathbf{x}}}, \quad (15.5)$$

$$\varphi_{\mathbf{grad}\,\vartheta} = 0, \quad (15.6)$$

$$\mathbf{P} = -\rho\varphi_{\boldsymbol{\mathcal{E}}}, \quad (15.7)$$

$$\boldsymbol{\mathcal{M}} = -\rho\varphi_{\mathbf{B}}, \quad (15.8)$$

$$\tau = -\rho^2\varphi_\rho I, \quad (15.9)$$

$$\boldsymbol{\mathcal{J}} \cdot \boldsymbol{\mathcal{E}} - (\mathbf{q} \cdot \mathbf{grad}\,\vartheta)/\vartheta \geq 0. \quad (15.10)$$

Equation (15.6) shows that φ cannot depend on the temperature gradient. Hence it is (at most) a function $\varphi(\rho, \vartheta, \dot{\mathbf{x}}, \boldsymbol{\mathcal{E}}, \mathbf{B})$ of the eleven variables ρ, ϑ, $\dot{\mathbf{x}}$, $\boldsymbol{\mathcal{E}}$ and $\mathbf{B}$. According to (15.4), the specific entropy cannot be given by an arbitrary constitutive relation, as we have assumed above for the twenty-six quantities: it must be equal to $-\varphi_\vartheta$; in particular it, too, is independent of $\mathbf{grad}\,\vartheta$. Similar remarks hold for $\mathbf{g}$, $\mathbf{P}$, $\mathcal{M}$, T (cf. (14.12)) and ε (cf. (14.10)). This is not all. For if $\mathbf{g}$ is to reduce to $\dot{\mathbf{x}}$ in the absence of an electromagnetic field, we must have $\varphi_{\dot{\mathbf{x}}}(\rho, \vartheta, \dot{\mathbf{x}}, 0, 0) = 0$. According to (15.7) and (15.8), $\mathbf{P}$ and $\mathcal{M}$ are functions of ρ, ϑ, $\dot{\mathbf{x}}$, $\mathcal{E}$ and $\mathbf{B}$. Now $\mathbf{P}$ and $\mathcal{M}$ are Galilei invariants (cf. §9), as are their arguments ρ, ϑ, $\mathcal{E}$ and $\mathbf{B}$, but *not* the argument $\dot{\mathbf{x}}$. Thus $\varphi_{\mathcal{E}}$ and $\varphi_{\mathbf{B}}$ must be independent of $\dot{\mathbf{x}}$, which is the same as saying that the mixed second partial derivatives $\varphi_{\mathcal{E}\dot{\mathbf{x}}}$ and $\varphi_{\mathbf{B}\dot{\mathbf{x}}}$ must vanish. Hence $\varphi_{\dot{\mathbf{x}}}$ is independent of either $\mathcal{E}$ or $\mathbf{B}$, and the foregoing requirement that $\varphi_{\dot{\mathbf{x}}}(\rho, \vartheta, \dot{\mathbf{x}}, 0, 0) = 0$ leads to the conclusion that $\varphi_{\dot{\mathbf{x}}} = 0$ *always*. The variables on which φ may depend are thus reduced by another three.

We now have

$$\eta = -\varphi_\vartheta(\rho, \vartheta, \mathcal{E}, \mathbf{B}), \tag{15.11}$$

$$\mathbf{g} = \dot{\mathbf{x}} + \epsilon_0 \mathbf{E} \times \mathbf{B}/\rho, \tag{15.12}$$

$$\mathbf{P} = -\rho\varphi_{\mathcal{E}}, \tag{15.13}$$

$$\mathcal{M} = -\rho\varphi_{\mathbf{B}}, \tag{15.14}$$

$$T = -[\rho^2\varphi_\rho + \epsilon_0 E^2/2 + B^2/(2\mu_0) - \mathcal{M}\cdot\mathbf{B}]I$$

$$+\epsilon_0 \mathbf{E} \otimes \mathbf{E} + \mathbf{B} \otimes \mathbf{B}/\mu_0 + \mathcal{E} \otimes \mathbf{P} - \mathcal{M} \otimes \mathbf{B} + \epsilon_0 \mathbf{E} \times \mathbf{B} \otimes \dot{\mathbf{x}}, \quad (15.15)$$

and the remnant (15.10) of the entropy inequality, with which we shall deal presently. Since the entropy, the stress, the polarization and the magnetization are all determined by derivatives of the free energy, the latter deserves to be called a *thermodynamic potential*.

We must still impose Euler's second law of mechanics (12.11) as a *second* constraint on the constitutive relations. If we substitute (15.12)–(15.15) in (12.11), we obtain the conditions

$$\mathcal{E}_i\varphi_{\mathcal{E}_j} + B_i\varphi_{B_j} = \mathcal{E}_j\varphi_{\mathcal{E}_i} + B_j\varphi_{B_i}. \tag{15.16}$$

PROBLEM

Problem 15–1. If $f(\mathbf{a})$ is differentiable, prove that $f(Q\mathbf{a}) = f(\mathbf{a})$ for every orthogonal matrix Q if, and only if,

$$a_i \frac{\partial f}{\partial a_j} = a_j \frac{\partial f}{\partial a_i} \qquad (15.17)$$

for all i and j.

The conditions (15.16) restrict the way in which φ depends on the vectors $\boldsymbol{\mathcal{E}}$ and $\mathbf{B}$: it must be invariant with respect to rotations. For example, φ cannot be a function of the sum of the Cartesian components of $\mathbf{B}$.

The original list of twenty-six functions of fourteen variables has been narrowed down to six ($\boldsymbol{\mathcal{J}}$ and $\mathbf{q}$), plus one function φ of eight variables only. This is obviously a considerable reduction, but it is not complete, because the constitutive relations for the conduction current density $\boldsymbol{\mathcal{J}}$ and the heat flux $\mathbf{q}$ must still satisfy the inequality (15.10), which is all that survives from the original entropy inequality. For an *insulator* (with respect to *both* electricity and heat) these constitutive relations are $\boldsymbol{\mathcal{J}} = \mathbf{q} = 0$ by definition, and the inequality (15.10) reduces to the *equality* $0 = 0$. We conclude that, among the materials in the class under consideration, *the insulators are incapable of undergoing irreversible processes*. We shall discuss *conductors* in the next section. But we note that, even for conductors, all processes throughout which $\boldsymbol{\mathcal{E}} = \mathbf{grad}\,\vartheta = 0$ are reversible. There may, of course, be other processes for which (15.10) becomes an equality.

The set of values which the arguments of the constitutive relations assume throughout a body at a given instant is called a *state* of the body. We shall refer to the states with $\boldsymbol{\mathcal{E}} = \mathbf{grad}\,\vartheta = 0$ as *special*, and to the processes that are successions of special states as *special processes*. Although the special states do not imply any balance of forces or torques, it is clear that they, together with the special processes – which need not be 'slow' – have the essential properties that classical thermodynamics associates with 'equilibrium' and 'quasi-static process'.

Consider now equation (15.15) for the stress tensor. We shall use the common notation

$$p = \rho^2 \varphi_\rho(\rho, \vartheta, \boldsymbol{\mathcal{E}}, \mathbf{B}); \qquad (15.18)$$

this is called the *pressure*.

PROBLEMS

Problem 15–2. Prove the identity

$$\mathbf{div}\,(\mathbf{E} \otimes \mathbf{E} - \tfrac{1}{2}E^2 I) = \mathbf{E}\,\mathrm{div}\,\mathbf{E} - \mathbf{E} \times \mathbf{curl}\,\mathbf{E}. \qquad (15.19)$$

Problem 15–3. With the aid of Maxwell's equations, prove that, for the T of (15.15),

$$\mathrm{div}\,T = -\,\mathbf{grad}\,p + (\,\mathcal{M} \cdot \mathbf{grad})\mathbf{B} + \mathcal{M} \times \mathbf{curl}\,\mathbf{B} + (\mathbf{P} \cdot \mathbf{grad})\mathcal{E}$$

$$+q\mathbf{E} + (\mathbf{j} + \overset{*}{\mathbf{P}}) \times \mathbf{B} + \rho(\epsilon_0 \mathbf{E} \times \mathbf{B}/\rho)^{\cdot}. \qquad (15.20)$$

Problem 15–4. Verify that

$$q\mathbf{E} + \mathbf{j} \times \mathbf{B} = q\mathcal{E} + \mathcal{J} \times \mathbf{B}. \qquad (15.21)$$

If we substitute (15.12) and (15.20)–(15.21) in (12.10), we obtain Euler's first law in the form

$$\rho\ddot{\mathbf{x}} = -\,\mathbf{grad}\,p + q\mathcal{E} + \mathcal{J} \times \mathbf{B} + (\mathbf{P} \cdot \mathbf{grad})\mathcal{E}$$

$$+\overset{*}{\mathbf{P}} \times \mathbf{B} + (\,\mathcal{M} \cdot \mathbf{grad})\mathbf{B} + \mathcal{M} \times \mathbf{curl}\,\mathbf{B} + \rho\mathbf{b}. \qquad (15.22)$$

We note that the electromagnetic part $\epsilon_0 \mathbf{E} \times \mathbf{B}/\rho$ of $\mathbf{g}$ has dropped out. It should, perhaps, be emphasized that, unless $\mathbf{P}$ and $\mathcal{M}$ both vanish, the pressure p depends on the electromagnetic field, and the first term on the right hand side of (15.22) is as 'electromagnetic' as the others. Another remarkable fact worth noting is that (15.22) is a Galilei-invariant equation.

PROBLEM

Problem 15–5. Show that the energy balance law (12.12) can be written in the form

$$\rho\vartheta\dot{\eta} = \mathcal{J} \cdot \mathcal{E} + \rho h - \mathrm{div}\,\mathbf{q}. \qquad (15.23)$$

Despite its appearance, this is the first principle of thermodynamics for materials in the class under consideration. It is *not* the second principle for reversible processes.

To sum up the results of this lengthy discussion, a material of the class we have been considering is determined by two vectors $\mathcal{J}$ and $\mathbf{q}$, which depend on the arguments $(\rho, \vartheta, \mathcal{E}, \mathbf{B}, \mathbf{grad}\, \vartheta)$ and satisfy the inequality (15.10), and by a function $\varphi(\rho, \vartheta, \mathcal{E}, \mathbf{B})$, which must be invariant under orthogonal transformations of its arguments $\mathcal{E}$ and $\mathbf{B}$. The entropy, momentum, polarization, magnetization and stress are determined by (15.11)–(15.15). The behaviour of any such material is governed by the law of mass conservation, by (15.22)–(15.23) and by Maxwell's equations together with the aether relations. Solutions of these equations need not be checked for compliance with Euler's second law of mechanics or with the second principle of thermodynamics, because these laws have already 'been taken care of'.

Since the whole analysis concerned the local laws, the results we have found apply on either side of any surface of discontinuity, irrespective of whether or not the surface is a boundary between two materials with different constitutive relations. Across any such surface, *all* the jump conditions must be satisfied. They may affect the speed at which the surface is moving (as was the case when we applied the electromagnetic jump conditions across a shock), its location and shape, and even the solutions far away from the surface.

For one special class of materials we have succeeded in setting up an Electrodynamics of Moving Bodies.† It is, of course, a non-relativistic theory, and is therefore suitable only for slowly moving bodies (with velocities, relative to an aether frame, not exceeding thirty million km/h, say). It is indeed possible to formulate relativistic analogues of this theory, but they are not definitive, because an accepted relativistic thermodynamics does not exist. Several relativistic theories of mechanics and thermodynamics have in fact been proposed, but the absence – so far – of any clear experimental evidence has made it impossible to decide which one of them, if any, should be preferred. The one feature they all share – besides Lorenz invariance – is that they reduce to the classical set of mechanical and thermodynamical laws when $\dot{x}^2$ is everywhere small compared with c^2. It is important to note that $\dot{\mathbf{x}}$ here means the velocity relative to any given Lorenz (or aether) frame. So long as we

† This is the title of Einstein's famous paper of 1905.

limit ourselves to systems of bodies in which the relative velocities are all small compared to c, we shall always be able to choose an aether (or Lorenz) frame relative to which $\dot{x}^2 \ll c^2$ everywhere. In this aether frame every relativistic theory will reduce to the classical laws (12.9)–(12.19). It is in this sense, and under these limited circumstances, that we shall henceforth apply the classical laws in *an* aether frame (rather than *the* aether frame). So far as terrestrial experiments are concerned, the foregoing limitation is not serious, because macroscopic bodies cannot be accelerated to relativistic speeds. It is only in astronomy that we find evidence for relative velocities approaching the velocity of light. For charged particles, on the other hand, a relativistic theory is both necessary and available, but it is not the same as the one we have developed for macroscopic (often called *ponderable*) bodies. After all, a particle has no energy (in the thermodynamic sense), temperature or entropy. We shall deal with charged particles in a later section.

Of course, Coleman and Noll's method can, and indeed must, be applied separately for each class of materials, that is, for each class of constitutive relations. We must not expect the terms on the right hand side of the 'equation of motion' (15.22) to be the same for all bodies. There is no *universal* or *standard* expression for the electromagnetic force on a body (except in the case of charged particles, where a universal expression is laid down as an axiom). In this sense, electromagnetism is really a theory of materials. We shall treat other classes of materials in the following sections.

As for the materials of the class with which we have dealt in this section, the subclass of those that are neither polarizable nor magnetizable is especially simple: according to (15.13) and (15.14), they are characterized by a free energy that depends only on ρ and ϑ. Such materials, sometimes called *pure conductors*, furnish the subject matter of *magnetohydrodynamics*. For them equation (15.22) becomes

$$\rho\ddot{\mathbf{x}} = -\operatorname{\mathbf{grad}} p(\rho, \vartheta) + q\boldsymbol{\mathcal{E}} + \boldsymbol{\mathcal{J}} \times \mathbf{B} + \rho\mathbf{b}$$

$$= -\operatorname{\mathbf{grad}} p + q\mathbf{E} + \mathbf{j} \times \mathbf{B} + \rho\mathbf{b}. \tag{15.24}$$

This equation is simple enough to serve as the basis for discussing experiments designed to fix the electromagnetic units. For this subclass of materials, the $\mathbf{grad}\, p(\rho, \vartheta)$ term is purely 'thermodynamic', because it does not involve $\mathbf{E}$ or $\mathbf{B}$. Similarly (again, only for this subclass), the energy,

$$\varepsilon = \varphi(\rho, \vartheta) - \vartheta\varphi_\vartheta(\rho, \vartheta) + \tfrac{1}{2}\dot{x}^2 + [\epsilon_0 E^2/2 + B^2/(2\mu_0)]/\rho, \qquad (15.25)$$

is the sum of parts which may be called thermodynamic, kinetic, electric and magnetic. There is no harm in using these names for the various terms in (15.25). But there is the real danger that usage may lead to abusage: for example, one might forget that the statement '$\epsilon_0 E^2/2$ is the electric energy density' is just the consequence of a particularly simple constitutive relation, and then proceed to apply it to polarizable materials – with disastrous effects.

Finally, for these pure conductors, the stress is given by

$$T = -[p(\rho, \vartheta) + \epsilon_0 E^2/2 + B^2/(2\mu_0)]I$$

$$+\epsilon_0 \mathbf{E} \otimes \mathbf{E} + \mathbf{B} \otimes \mathbf{B}/\mu_0 + \epsilon_0 \mathbf{E} \times \mathbf{B} \otimes \dot{\mathbf{x}}. \qquad (15.26)$$

The complete stress tensor – effectively, the function $p(\rho, \vartheta)$, since everything else appears explicitly in the foregoing expression – can be determined for these materials from measurements carried out under very special circumstances: the material may be stationary, the temperature uniform, and the electromagnetic field absent altogether. This was not at all obvious when we began the discussion by assuming constitutive relations of the form (15.1) for the components of the stress tensor.

Another useful subclass is that of *incompressible* fluids. These fluids have a constant density, which may be omitted from the constitutive relations. By the law of mass conservation, their velocity fields are constrained to be volume-preserving: $\mathrm{div}\, \dot{\mathbf{x}} = 0$. This has the following consequences: first, the terms involving φ_ρ drop out of (15.2) and (15.3), and the sixth term in (15.3) becomes simply $\tau \cdot \mathrm{grad}\, \dot{\mathbf{x}}$. This scalar product of the matrices τ and $\mathrm{grad}\, \dot{\mathbf{x}}$ has to satisfy the entropy inequality, but now $\mathrm{grad}\, \dot{\mathbf{x}}$ is not arbitrary, because $\mathrm{div}\, \dot{\mathbf{x}}$ must vanish. We note, however, that $\mathrm{div}\, \dot{\mathbf{x}} = I \cdot \mathrm{grad}\, \dot{\mathbf{x}}$. The requirement that $\tau \cdot \mathrm{grad}\, \dot{\mathbf{x}} = 0$ for all $\mathrm{grad}\, \dot{\mathbf{x}}$ such that $I \cdot \mathrm{grad}\, \dot{\mathbf{x}} = 0$ therefore means that τ must be perpendicular to every matrix that is perpendicular to the unit matrix.

It follows that τ must be parallel to I:

$$\tau = -pI. \tag{15.27}$$

This relation, in which the 'hydrostatic pressure' p is now an arbitrary number (at each event in space-time), replaces (15.9). All subsequent relations are unchanged if we replace $\rho^2\varphi_\rho$ everywhere by p and remember that ρ is a given constant. In particular, (15.22) is unchanged, and in fact serves to determine $p(\mathbf{x}, t)$. In practice, we take the **curl** of (15.22), which obliterates $\mathbf{grad}\, p$, and solve the resulting equation, together with $\operatorname{div} \dot{\mathbf{x}} = 0$; *then* we use (15.22) as a formula for $\mathbf{grad}\, p$; this determines p to within an additive constant; and the latter is usually fixed by the jump conditions.

It should, perhaps, be noted that the pressure p in an incompressible fluid will generally not be the same as in a *compressible* fluid that is happening to undergo a volume-preserving motion, even if it is the *same* motion. In the latter case p must equal $\rho^2\varphi_\rho$, and in the resulting solution ρ will generally fail to be a constant.

16. Elastic materials

Materials with constitutive relations of the form

$$f = \mathcal{F}(F, \vartheta, \dot{\mathbf{x}}, \boldsymbol{\mathcal{E}}, \mathbf{B}, \mathbf{grad}\,\vartheta), \tag{16.1}$$

where F is the deformation tensor (11.2), are called *elastic*. They are often used as prototypes of solid materials, but this must not be taken too literally. In fact, the remarks following (15.1) apply here as well. The fluids defined by (15.1) are a subclass of the elastic materials; they depend on the components of F through $\rho \propto (\det F)^{-1}$. Frankly, the reason for calling these materials elastic is that $f = \mathcal{F}(F, \vartheta, \mathbf{grad}\,\vartheta)$ is the constitutive relation in most studies devoted to the theory of elasticity. We simply extend this to electromagnetism by adding $\dot{\mathbf{x}}$, the velocity with respect to the aether frame we have chosen, and the electromagnetic vectors $\boldsymbol{\mathcal{E}}$ and $\mathbf{B}$.

If we now try to follow the analysis of the previous section with the new kind of constitutive relation, we immediately see that the term $\varphi_\rho\dot{\rho}$ in (15.2) must be replaced by the double sum $(\partial\varphi/\partial F_{ij})\dot{F}_{ij}$. It is convenient to define a *matrix* φ_F by $(\varphi_F)_{ij} = \partial\varphi/\partial F_{ij}$, so that this sum becomes the matrix scalar product $\varphi_F \cdot F$. This causes the replacement, in the entropy inequality (15.3), of the term $(\tau + \rho^2\varphi_\rho I) \cdot \operatorname{grad} \dot{\mathbf{x}}$ by

$$\tau \cdot \operatorname{grad} \dot{\mathbf{x}} - \rho\varphi_F \cdot \dot{F}.$$

PROBLEM

Problem 16–1. Use (11.10) and $\operatorname{tr} BA = \operatorname{tr} AB$ to show that

$$\tau \cdot \operatorname{grad} \dot{\mathbf{x}} - \rho\varphi_F \cdot \dot{F} = [\tau(F^T)^{-1} - \rho\varphi_F] \cdot \dot{F}.$$

From the last result it follows that (15.9) is replaced by $\tau = \rho\varphi_F F^T$. In fact, the whole argument leading up to (15.11)–(15.15) goes through with this single change, and we obtain the relations

$$\eta = -\varphi_\vartheta(F, \vartheta, \boldsymbol{\mathcal{E}}, \mathbf{B}), \tag{16.2}$$

$$\mathbf{g} = \dot{\mathbf{x}} + \epsilon_0 \mathbf{E} \times \mathbf{B}/\rho, \tag{16.3}$$

$$\mathbf{P} = -\rho\varphi_{\mathcal{E}}, \tag{16.4}$$

$$\boldsymbol{\mathcal{M}} = -\rho\varphi_{\mathbf{B}}, \tag{16.5}$$

$$T = \rho\varphi_F F^T - [\epsilon_0 E^2/2 + B^2/(2\mu_0) - \boldsymbol{\mathcal{M}} \cdot \mathbf{B}]I$$

$$+\epsilon_0 \mathbf{E} \otimes \mathbf{E} + \mathbf{B} \otimes \mathbf{B}/\mu_0 + \boldsymbol{\mathcal{E}} \otimes \mathbf{P} - \boldsymbol{\mathcal{M}} \otimes \mathbf{B} + \epsilon_0 \mathbf{E} \times \mathbf{B} \otimes \dot{\mathbf{x}}, \tag{16.6}$$

$$\boldsymbol{\mathcal{J}} \cdot \boldsymbol{\mathcal{E}} - (\mathbf{q} \cdot \operatorname{grad} \vartheta)/\vartheta \geq 0. \tag{16.7}$$

It should, perhaps, be noted that the partial derivative φ_ϑ in (16.2) is with F held constant, whereas in the corresponding equation (15.11) it was taken with ρ (or $\det F$) held constant. The same remark applies to $\varphi_{\mathcal{E}}$ and $\varphi_{\mathbf{B}}$.

From the remnant (16.7) of the entropy inequality we conclude that, in elastic materials, as in fluids, all special processes throughout which $\boldsymbol{\mathcal{E}} = \operatorname{grad} \vartheta = 0$ are reversible. Of course, (16.7) may also become an equality for other kinds of processes.

Equations (15.22)–(15.26) of the previous section will hold for elastic materials if $-pI$ is replaced by $\rho\varphi_F F^T$, and $-\operatorname{\mathbf{grad}} p = \operatorname{\mathbf{div}}(-pI)$ by $\operatorname{\mathbf{div}} \rho\varphi_F F^T$.

As an application, we consider two straight and parallel wires carrying currents i and i'. Neither wire is moving (with respect to an aether frame). If one of them consists of purely conducting elastic material, we shall have

$$0 = \operatorname{\mathbf{div}} \rho\varphi_F F^T + q\mathbf{E} + \mathbf{j} \times \mathbf{B} + \rho\mathbf{b} \qquad (16.8)$$

holding inside it. We are interested in $\mathbf{j} \times \mathbf{B}$, but in this equation it is balanced by three other terms. The term $q\mathbf{E}$ will vanish if the charge density vanishes. The first term will usually not vanish, but its integral over the wire may be expected to be small: by the jump conditions, this integral can be shown to be equal to the pressure force of the surrounding air. In this way $\int \mathbf{j} \times \mathbf{B}\, dV$ may be measured by balancing it against gravity, $\int \mathbf{b}\, dm$ (with $\mathbf{b}$ equal to the acceleration of gravity). Now, in $\int \mathbf{j} \times \mathbf{B}\, dV$ there is the contribution of the self-field $\mathbf{B}$ due to the current density $\mathbf{j}$ flowing in the wire. If the distribution of $\mathbf{j}$ is symmetrical, this contribution vanishes and, by (8.10), the force $\int \mathbf{j} \times \mathbf{B}\, dV$ becomes $\mu_0 ii'/(2\pi r)$ per unit length of the wire, where r is the distance between the wires. It is perpendicular to the wire and directed towards (away from) the other wire if the currents are parallel (anti-parallel). If $i = i' = 1$ amp, $r = 1$ m and we use (12.7), the force is 2×10^{-7} newton per metre of the wire. This arrangement and this result are often cited as the definition of the ampere: 'The *ampere* is the steady current that, if flowing in two parallel thin conductors of infinite length and one metre apart, would produce a force of 2×10^{-7} newton per metre between them'. Since the experiment, like any experiment, is fraught with uncertainties and is, moreover, based on special constitutive assumptions, it is better to regard it as an attempt to verify a special force formula that follows from the definition (12.7) of the ampere. That definition, we recall, was independent of the constitution of any material. (We might say that it was based on the constitution of the *aether*, as expressed by the universal aether relations.)

When we now impose Euler's second law of mechanics as a further restriction on the constitutive relations, we find that the free energy must be such that

$$\varphi(QF, \vartheta, Q\boldsymbol{\mathcal{E}}, Q\mathbf{B}) = \varphi(F, \vartheta, \boldsymbol{\mathcal{E}}, \mathbf{B}) \tag{16.9}$$

holds for every orthogonal matrix Q. Having chosen some function $\varphi(F, \vartheta, \boldsymbol{\mathcal{E}}, \mathbf{B})$ to describe any particular elastic material, it may seem that we are in the awkward position of having to verify (16.9) at each material point at every instant. Fortunately, it is possible to take care of this restriction by using an algebraic theorem of Cauchy: any non-singular matrix F has a unique decomposition of the form

$$F = RU, \tag{16.10}$$

where R is orthogonal and U is a symmetric, positive-definite matrix. We use this decomposition for the deformation F. In (16.9) we may choose the orthogonal matrix R^T for Q. Then $QF = R^T RU = U$, $Q\boldsymbol{\mathcal{E}} = R^T \boldsymbol{\mathcal{E}} = U^{-1} U R^T \boldsymbol{\mathcal{E}} = U^{-1} F^T \boldsymbol{\mathcal{E}}$ and, similarly, $Q\mathbf{B} = U^{-1} F^T \mathbf{B}$. Hence

$$\varphi(F, \vartheta, \boldsymbol{\mathcal{E}}, \mathbf{B}) = \varphi(U, \vartheta, U^{-1} F^T \boldsymbol{\mathcal{E}}, U^{-1} F^T \mathbf{B}).$$

But $F^T F = (RU)^T RU = U^T U = U^2$. It follows that φ is a function of $F^T F$, ϑ, $F^T \boldsymbol{\mathcal{E}}$ and $F^T \mathbf{B}$:

$$\varphi = \Phi(F^T F, \vartheta, F^T \boldsymbol{\mathcal{E}}, F^T \mathbf{B}). \tag{16.11}$$

Conversely, if φ is *any* function of the form (16.11), it automatically satisfies the requirement (16.9) for *any* orthogonal Q. We shall therefore refer to (16.11) as the free energy in *reduced* form. Whereas $\varphi(F, \vartheta, \boldsymbol{\mathcal{E}}, \mathbf{B})$ depends on the nine components of the deformation matrix F, the reduced Φ depends only on the six independent components of the *symmetric* matrix $F^T F$. This reduction in the number of arguments is, of course, a direct result of imposing the requirements expressed by the three components of Euler's second law of mechanics. We shall see (in Chapter VII) how the simplest (linear) form of the reduced free energy can be used to predict several important effects in dielectrics.

Next, let us discuss *conductors*, that is, materials in which $\boldsymbol{\mathcal{J}}$ and $\mathbf{q}$ do not vanish. In the first place, the Galilei invariants $\boldsymbol{\mathcal{J}}$ and $\mathbf{q}$ (like $\mathbf{P}$ and $\boldsymbol{\mathcal{M}}$) cannot depend on $\dot{\mathbf{x}}$, along with the other Galilei-invariant arguments F (or $\rho \propto (\det F)^{-1}$ for a fluid), ϑ, $\boldsymbol{\mathcal{E}}$, $\mathbf{B}$ and $\mathbf{grad}\,\vartheta$. Unlike the other quantities, however, there is no reason why $\mathcal{J}$ and $\mathbf{q}$ should not depend on $\mathbf{grad}\,\vartheta$. We can establish an important property of these two vectors by considering, for arbitrary F, ϑ, $\mathbf{B}$, $\mathbf{a}$ and $\mathbf{b}$, the function

$$\gamma(\lambda) = \mathcal{J}(F,\vartheta,\lambda\mathbf{a},\mathbf{B},\lambda\mathbf{b}) \cdot \lambda\mathbf{a} - \mathbf{q}(F,\vartheta,\lambda\mathbf{a},\mathbf{B},\lambda\mathbf{b}) \cdot \lambda\mathbf{b}/\vartheta. \qquad (16.12)$$

According to (15.10) or (16.7), $\gamma \geq 0$ always. Evidently $\gamma(0) = 0$, which means that the lower bound, 0, of $\gamma(\lambda)$ is actually a *minimum* at $\lambda = 0$. If $\mathcal{J}$ and $\mathbf{q}$ are differentiable, it follows that $\gamma'(0) = 0$ (a necessary, but not a sufficient, condition for a minimum). According to (16.12), this condition requires

$$\mathcal{J}(F,\vartheta,0,\mathbf{B},0) \cdot \mathbf{a} - \mathbf{q}(F,\vartheta,0,\mathbf{B},0) \cdot \mathbf{b} = 0 \qquad (16.13)$$

for arbitrary F, ϑ, $\mathbf{B}$, $\mathbf{a}$ and $\mathbf{b}$. Hence

$$\mathcal{J}(F,\vartheta,0,\mathbf{B},0) = \mathbf{q}(F,\vartheta,0,\mathbf{B},0) = 0, \qquad (16.14)$$

which means that $\mathcal{J}$ and $\mathbf{q}$ must both vanish whenever the electromotive intensity $\mathcal{E}$ and the temperature gradient both vanish, i.e. whenever the state is a special one. We emphasize that this is a corollary, arrived at by Coleman and Noll's method. We have certainly not assumed Ohm's law $\mathcal{J} = \sigma\mathcal{E}$ or Fourier's law $\mathbf{q} = -\kappa\,\mathbf{grad}\,\vartheta$. (In classical thermodynamics 'equilibrium' is defined as a state in which, among other things, $\mathbf{grad}\,\vartheta$ and $\mathbf{q}$ *both* vanish.)

The vectors $\mathcal{J}$ and $\mathbf{q}$ now depend on F, ϑ, $\mathcal{E}$, $\mathbf{B}$ and $\mathbf{grad}\,\vartheta$. We have just shown that they both vanish whenever $\mathcal{E} = \mathbf{grad}\,\vartheta = 0$, but this corollary does not exhaust the implications of the inequality (16.7), because it is a necessary, not a sufficient, condition for the inequality to hold. For the subclass of *linearly conducting materials*, $\mathcal{J}$ and $\mathbf{q}$ are assumed to be linear in $\mathcal{E}$ and $\mathbf{grad}\,\vartheta$. By the corollary, these linear relations must be homogeneous:

$$\mathcal{J} = a\mathcal{E} + b\,\mathbf{grad}\,\vartheta,$$

$$\mathbf{q} = c\mathcal{E} + d\,\mathbf{grad}\,\vartheta, \qquad (16.15)$$

where a, b, c and d are matrices with elements depending on F, ϑ and $\mathbf{B}$.

17. Viscous materials

Viscous materials are those in which velocity gradients matter, so that the nine components of the velocity gradient $\operatorname{grad} \dot{x}$ must be added as further arguments in the constitutive relations. We shall restrict ourselves to a brief discussion of the consequences that follow from this addition in the case of fluids.

Starting from the entropy inequality (14.11), it can easily be shown that φ does not depend on *either* $\operatorname{grad} \vartheta$ or $\operatorname{grad} \dot{x}$. The relations (15.4)–(15.8) still hold. Hence ε, η, $\mathbf{g}$, $\mathbf{P}$ and $\mathcal{M}$, too, are independent of either $\operatorname{grad} \vartheta$ or $\operatorname{grad} \dot{x}$. But since the τ of (14.12) now depends on $\operatorname{grad} \dot{x}$, $\tau + pI$ no longer vanishes. In fact, it will now join $\mathcal{J}$ and $\mathbf{q}$ in the inequality

$$(\tau + pI) \cdot \operatorname{grad} \dot{x} + \mathcal{J} \cdot \mathcal{E} - (\mathbf{q} \cdot \operatorname{grad} \vartheta)/\vartheta \geq 0, \qquad (17.1)$$

which replaces (15.9)–(15.10). It follows that, in viscous fluids, all processes throughout which $\mathcal{E}$, $\operatorname{grad} \vartheta$ and $\operatorname{grad} \dot{x}$ all vanish are reversible. Of course, there may be other processes for which (17.1) becomes an equality.

By a direct generalization of the argument that has led to (16.14), we arrive at the following result: whenever $\operatorname{grad} \dot{x}$, $\mathcal{E}$ and $\operatorname{grad} \vartheta$ all vanish, so do $\tau + pI$, $\mathcal{J}$ and $\mathbf{q}$; in particular, the stress T will then be given by (15.15). For viscous fluids, the special states are obviously those in which the temperature and the velocity are both uniform; moreover, the motion and the electromagnetic field are everywhere constrained by $\mathbf{E} + \dot{x} \times \mathbf{B} = 0$. The special processes that are successions of these special states have the essential properties that classical thermodynamics associates with 'quasi-static processes'. For example, along a special process, $\Delta S = \int (\mathcal{Q}/\vartheta) \, dt$.

In *linearly viscous* fluids, $\tau + pI$, $\mathcal{J}$ and $\mathbf{q}$ are assumed to be linear in $\operatorname{grad} \dot{x}$, $\mathcal{E}$ and $\operatorname{grad} \vartheta$. By the foregoing result, these linear relations must be homogeneous. In isotropic materials, they become especially simple: the matrix $\tau + pI$ cannot depend on either of the vectors $\mathcal{E}$ or $\operatorname{grad} \vartheta$, and the vectors $\mathcal{J}$ and $\mathbf{q}$ cannot depend on the matrix $\operatorname{grad} \dot{x}$. For $\mathcal{J}$ and $\mathbf{q}$ we are, again, led to (16.15). For $\tau + pI$, we obtain a linear relation of the form

$$\tau + pI = 2\lambda d + \zeta(\operatorname{div} \dot{\mathbf{x}})I + 2\nu e, \qquad (17.2)$$

where

$$2d = \operatorname{grad} \dot{\mathbf{x}} + (\operatorname{grad} \dot{\mathbf{x}})^T, \qquad 2e = \operatorname{grad} \dot{\mathbf{x}} - (\operatorname{grad} \dot{\mathbf{x}})^T, \qquad (17.3)$$

and λ, ζ and ν are scalar coefficients. Finally, it follows from (17.1) that ν must vanish, whereas λ and ζ are restricted by the *Navier-Stokes relations*

$$\lambda \geq 0, \qquad 3\zeta + 2\lambda \geq 0. \qquad (17.4)$$

PROBLEMS

Problem 17-1. Show that

$$\operatorname{div}\left[2\lambda d + \zeta(\operatorname{div} \dot{\mathbf{x}})I\right] = (2\lambda + \zeta)\, \mathbf{grad}\, \operatorname{div} \dot{\mathbf{x}} - \lambda \, \mathbf{curl}^2 \dot{\mathbf{x}}. \qquad (17.5)$$

Problem 17-2. Show that, for linearly viscous fluids, the law of energy balance (cf. (15.23)) becomes

$$\rho \vartheta \dot{\eta} = \boldsymbol{\mathcal{J}} \cdot \boldsymbol{\mathcal{E}} + 2\lambda \operatorname{tr}(d \cdot d) + \zeta(\operatorname{tr} d)^2 + \rho h - \operatorname{div} \mathbf{q}. \qquad (17.6)$$

18. Charged particles

A classical charged particle is a specially simple electromagnetic material: it is a point with which we associate a positive mass m and a charge e – positive or negative – but no conduction current, polarization, magnetization, energy, temperature or entropy. From the point of view of the general theory we have developed for continuous bodies it is so special as to be degenerate. It is not governed by the two principles of thermodynamics, because these cannot even be stated for a body that possesses neither energy, nor temperature or entropy. Euler's laws of mechanics are also given up in favor of an action principle. In fact, charged particles are governed by a theory which is entirely different from the one we have developed for continuous bodies. Whether we like it or not, it is an inescapable fact that, from a theoretical point of view, a macroscopic body is *not* an assembly of particles.

If we delete in equation (15.22) all terms connected with pressure, conduction current, polarization and magnetization, and then integrate over the volume of the body and pass to the limit of a particle with mass m and charge e, we obtain

$$m\ddot{\mathbf{x}} = e\boldsymbol{\mathcal{E}} + m\mathbf{b} = e(\mathbf{E} + \dot{\mathbf{x}} \times \mathbf{B}) + m\mathbf{b}. \tag{18.1}$$

But equation (15.22) has no meaning for a particle. If we therefore still wish to use (18.1) as an equation of motion for a charged particle, as we do, we must lay it down as an axiom. Like (15.22), it is a Galilei-invariant equation. In most of the applications of (18.1), the non-electromagnetic force $m\mathbf{b}$ is of little interest, either because it is small or because it is entirely absent. We shall therefore leave it out in the remainder of this section. The equation of motion is then simply

$$m\ddot{\mathbf{x}} = e(\mathbf{E} + \dot{\mathbf{x}} \times \mathbf{B}). \tag{18.2}$$

The expression on the right hand side of (18.2) is called *the Lorenz force*. If we were to include in it the fields produced by the particle itself, it would become infinite, and the equation of motion would not make any sense (at least, it would not make any classical sense; in quantum electrodynamics a particle is allowed to 'act on itself', subject to definite rules which govern the resulting infinities). We therefore stipulate that in the Lorenz force the self-fields are to be left out. Logically speaking, this is to be regarded as part of the definition of a classical charged particle. It should, perhaps, be noted that no such stipulation was necessary in the case of finite bodies.

From (18.2) we conclude that

$$\frac{d}{dt}\tfrac{1}{2}m\dot{x}^2 = e\dot{\mathbf{x}} \cdot \mathbf{E}. \tag{18.3}$$

The magnetic field can do no work on a particle because its contribution to the Lorenz force is normal to the path. If the electric field vanishes, the motion can still be very complicated, but the speed will be constant.

PROBLEM

Problem 18–1. A charged particle is projected into a uniform magnetic field **B**. If **B** is taken parallel to the z axis, show that the equations of motion are

$m\ddot{x} = eB\dot{y}$, $m\ddot{y} = -eB\dot{x}$, $m\ddot{z} = 0$. Deduce that the path of the particle is a helix described at constant speed.

We now seek a Lagrangian formulation for (18.2). The fields $\mathbf{E}$ and $\mathbf{B}$ are governed by the second principle of electromagnetism. In §6 we have shown that this can be ensured by expressing the fields in terms of potentials:

$$\mathbf{B} = \text{curl } \mathbf{A},$$

$$\mathbf{E} = -\mathbf{A}_t - \text{grad } V. \tag{18.4}$$

We also recall that $\mathbf{A}$ and V can be replaced by another pair, which describes the same $\mathbf{E}$ and $\mathbf{B}$, through the gauge transformation

$$\tilde{\mathbf{A}} = \mathbf{A} + \text{grad } \chi, \qquad \tilde{V} = V - \chi_t. \tag{18.5}$$

Now, if $\mathbf{A} = 0$, (18.2) becomes $m\ddot{\mathbf{x}} = -e\,\text{grad } V$, and a suitable Lagrangian for this case of motion under a conservative force would be $L = \frac{1}{2}m\dot{x}^2 - eV$. A gauge transformation will change this Lagrangian to $\tilde{L} = L + e\chi_t$. If the equation of motion is to be invariant with respect to gauge transformations – as it must, because $\mathbf{E}$ and $\mathbf{B}$ are unchanged by a gauge transformation – the difference $\tilde{L} - L$ must be (at most) a material time derivative; for that is the only change a Lagrangian may suffer without affecting Lagrange's equations. Thus $\dot{\chi} = \chi_t + \dot{\mathbf{x}} \cdot \text{grad } \chi$ would do, but χ_t does not. This leads us to propose

$$L(\mathbf{x}, \dot{\mathbf{x}}, t) = \frac{1}{2}m\dot{x}^2 + e[\dot{\mathbf{x}} \cdot \mathbf{A}(\mathbf{x}, t) - V(\mathbf{x}, t)] \tag{18.6}$$

as a Lagrangian function which, under a gauge transformation, changes to $\tilde{L} = L + e\dot{\chi}$.

PROBLEMS

Problem 18–2. Show that Lagrange's equations,

$$\frac{d}{dt}L_{\dot{\mathbf{x}}} - L_{\mathbf{x}} = 0, \tag{18.7}$$

with (18.6), lead to the equation of motion (18.2).

Problem 18–3. Prove that the four-dimensional contravariant vector $(v^\alpha) = (\dot{\mathbf{x}}, 1)$ is Galilei-invariant.

Problem 18–4. Use the fact that $(\mathbf{A}, -V)$ are the components of the electromagnetic four-potential V_α to prove that $\dot{\mathbf{x}} \cdot \mathbf{A} - V$ is Galilei-invariant.

If we define, for any path $\mathbf{x}(t)$, an *action* by

$$S = \int_{t_1}^{t_2} L\big(\mathbf{x}(t), \dot{\mathbf{x}}(t), t\big)\, dt, \tag{18.8}$$

then Lagrange's equations (18.7) are equivalent to the action principle

$$\delta S = 0, \tag{18.9}$$

where the variation is to be carried out with respect to all paths $\mathbf{x}(t)$ with given end points $\mathbf{x}_1 = \mathbf{x}(t_1)$ and $\mathbf{x}_2 = \mathbf{x}(t_2)$. For an assembly of particles, we take the action to be a sum of individual actions, one for each particle. By varying the action with respect to the path of the nth particle we are then led to the equation of motion (18.2) for that particle.

From the Lagrangian (18.6), it is a straightforward matter to pass to a Hamiltonian formulation. We define

$$\mathbf{p} = L_{\dot{\mathbf{x}}} = m\dot{\mathbf{x}} + e\mathbf{A} \tag{18.10}$$

as the generalized momentum and note that this relation may serve to express $\dot{\mathbf{x}}$ in terms of $\mathbf{p}$, $\mathbf{x}$ and t. We now form the Hamiltonian

$$H(\mathbf{p}, \mathbf{x}, t) = \mathbf{p} \cdot \dot{\mathbf{x}}(\mathbf{p}, \mathbf{x}, t) - L\big(\mathbf{x}, \dot{\mathbf{x}}(\mathbf{p}, \mathbf{x}, t), t\big). \tag{18.11}$$

PROBLEM

Problem 18–5. Show that

$$H(\mathbf{p}, \mathbf{x}, t) = \frac{(\mathbf{p} - e\mathbf{A})^2}{2m} + eV, \tag{18.12}$$

and that (18.10) and (18.2) follow from Hamilton's equations

$$\dot{\mathbf{x}} = H_{\mathbf{p}}, \qquad \dot{\mathbf{p}} = -H_{\mathbf{x}}. \tag{18.13}$$

From the Hamiltonian formulation, it is a straightforward matter to proceed to a formulation in terms of Poisson brackets. Such a canonical formulation is an indispensable prerequisite for setting up a quantum mechanics for charged particles. In fact, one of the objects of the classical theory of charged particles is to suggest a Hamiltonian for quantum mechanics. But the classical Hamiltonian formulation, too, has some advantages.

PROBLEM

Problem 18–6. Let (r, ϕ, z) be cylindrical coordinates. An electromagnetic field has rotational symmetry around the z axis if its components do not depend on ϕ. Prove that in any such field

$$m\dot{x}_\phi + eA_\phi = \gamma/r,$$

where γ is a constant of motion. If, furthermore, the electric field vanishes, prove that the available region for the particle is given by

$$|\gamma - erA_\phi(r, z)| \leq mr\sqrt{2m\mathcal{E}_k},$$

where $\mathcal{E}_k$ is the (constant) kinetic energy of the particle.

Since particles can be accelerated to relativistic speeds, the foregoing non-relativistic theory does not suffice. We indicate the changes that are necessary for rendering the formulation relativistic.

First, we take the action to be

$$S = \int [-mc^2 \sqrt{1 - \dot{x}^2/c^2} + e(\dot{\mathbf{x}} \cdot \mathbf{A} - V)] \, dt; \qquad (18.14)$$

the first term is now $- \int mc^2 \, d\tau$, where τ is the *proper time* along the path $\mathbf{x}(t)$; the second is $\int ev^\alpha V_\alpha \, d\tau$, where $(v^\alpha) = (\dot{\mathbf{x}}, 1)/\sqrt{1 - \dot{x}^2/c^2}$ is the velocity four-vector. The relativistic Lagrangian

$$L = -mc^2 \sqrt{1 - \dot{x}^2/c^2} + e(\dot{\mathbf{x}} \cdot \mathbf{A} - V) \qquad (18.15)$$

now leads to the generalized momentum

$$\mathbf{p} = \frac{m\dot{\mathbf{x}}}{\sqrt{1 - \dot{x}^2/c^2}} + e\mathbf{A} = \mathbf{p}_k + e\mathbf{A}, \qquad (18.16)$$

where the kinetic momentum

$$\mathbf{p}_k = \frac{m\dot{\mathbf{x}}}{\sqrt{1 - \dot{x}^2/c^2}} \tag{18.17}$$

replaces the former $m\dot{\mathbf{x}}$. Lagrange's equations with the Lagrangian (18.15) lead to the relativistic equation of motion

$$\dot{\mathbf{p}}_k = e(\mathbf{E} + \dot{\mathbf{x}} \times \mathbf{B}). \tag{18.18}$$

We define

$$\mathcal{E}_k = \frac{mc^2}{\sqrt{1 - \dot{x}^2/c^2}} \tag{18.19}$$

and call this *the relativistic kinetic energy*. For small speeds it reduces to $\frac{1}{2}m\dot{x}^2 + mc^2$. Thus $\mathcal{E}_k$ includes the constant *rest-energy* mc^2.

PROBLEM

Problem 18–7. Prove the relations

$$\mathbf{p}_k = \mathcal{E}_k \dot{\mathbf{x}}/c^2, \qquad \dot{\mathcal{E}}_k = \dot{\mathbf{x}} \cdot \dot{\mathbf{p}}_k. \tag{18.20}$$

From (18.18) and (18.20)$_2$ we obtain the relativistic analogue of (18.3):

$$\dot{\mathcal{E}}_k = e\dot{\mathbf{x}} \cdot \mathbf{E}. \tag{18.21}$$

It is still true that, if $\mathbf{E} = 0$, the speed is constant.

The relativistic Hamiltonian is obtained from (18.11), (18.15) and (18.16). A straightforward calculation leads to

$$H = \mathcal{E}_k + eV = \sqrt{m^2c^4 + (\mathbf{p} - e\mathbf{A})^2c^2} + eV. \tag{18.22}$$

We have used the equation of motion (15.22), which was derived by a simultaneous application of the laws of electromagnetism, mechanics and thermodynamics, in order to motivate our assumption of the Lorenz force. But an assumption, however motivated, is still an assumption, and it is important to realize that, once it has been made, there is no way back. It is impossible, without introducing further assumptions – such as are made in statistical mechanics – to start from equation (18.1) for a particle (or from the equivalent action principle) and arrive at a correct equation of motion for a macroscopic body by summing over many charged particles. Such a procedure would be tantamount to a derivation of thermodynamics from mechanics. It is for this reason that we stated, at the beginning of this section, that a macroscopic body cannot be simply regarded as an assembly of particles.

The foregoing remarks apply, as well, to the customary presentations of electromagnetism, in which the assumption of the Lorenz force on a particle is motivated differently (or not at all).

CHAPTER VI

Electrostatics

In the absence of any motion (through the aether) and any changes in time, Maxwell's equations (10.6) and the electromagnetic jump conditions are

$$\operatorname{div} \mathbf{D} = q, \qquad \mathbf{n} \cdot [\![\mathbf{D}]\!] = \sigma,$$

$$\operatorname{curl} \mathbf{E} = 0, \qquad \mathbf{n} \times [\![\mathbf{E}]\!] = 0,$$

$$\mathbf{D} = \epsilon_0 \mathbf{E} + \mathbf{P},$$

$$\operatorname{div} \mathbf{B} = 0, \qquad \mathbf{n} \cdot [\![\mathbf{B}]\!] = 0,$$

$$\operatorname{curl} \mathbf{H} = \mathbf{j}, \qquad \mathbf{n} \times [\![\mathbf{H}]\!] = \mathbf{K},$$

$$\mathbf{H} = \mathbf{B}/\mu_0 - \mathbf{M}.$$

It is obvious that any connection between the first and second halves of this system of equations can only be provided by $\mathbf{P}$, $\mathbf{j}$ or $\mathbf{M}$ (for a material at rest the Minkowsky $\mathbf{M}$ and the Lorenz $\mathcal{M}$ are the same; cf. $(9.5)_1$). Further discussion therefore depends on material properties. If $\mathbf{j} = \mathbf{M} = 0$ and $\mathbf{P}$ is independent of $\mathbf{B}$, the first half of the equations is independent of the second; its study is the subject of *electrostatics*. Analogously, if $\mathbf{j} = \mathbf{P} = 0$ and $\mathbf{M}$ is independent of $\mathbf{E}$, the second half is independent of the first; its study is the subject of *magnetostatics*. In spite of the similarity in the differential equations and aether relations, it is not true that every problem (with $q = 0$) in electrostatics translates into one in magnetostatics, and vice versa; for the jump conditions are different. But the mathematical techniques for solving problems in electrostatics and magnetostatics are quite similar. That is why only one of them, usually electrostatics, is studied in detail.

Finally, if the currents do not vanish, the whole set of equations forms the subject of *steady currents*.

19. Electrostatics of conductors

In electrostatics a conductor is a material that is not merely non-polarizable and non-magnetizable; it is also assumed to have the following constitutive property: any non-zero electromagnetic intensity $\mathcal{E}$ is associated with a non-zero conduction current $\mathcal{J}$. The material need not be a linear conductor in the sense of §16; in fact, we prefer not to commit ourselves to any definite constitutive class at this stage.

In the absence of motion $\mathcal{E} = \mathbf{E}$ and $\mathcal{J} = \mathbf{j}$, and we conclude that in electrostatics (i.e. $\mathbf{j} = 0$) the electric field $\mathbf{E}$ must vanish in the interior of any conductor. Furthermore, since $q = \operatorname{div} \epsilon_0 \mathbf{E} = 0$, there can be no charge density inside a conductor. Outside, where there is (by definition) an *insulator* (perhaps an insulating vacuum), $\mathbf{E}$ need not vanish, but the jump condition $\mathbf{n} \times [\![\mathbf{E}]\!] = 0$ must hold at the interface. Hence $\mathbf{E}$ on the outer side must be normal to the boundary. If we take the unit normal $\mathbf{n}$ to point out of the conductor, the jump conditions are

$$\mathbf{n} \times \mathbf{E} = 0,$$

$$D_n = \epsilon_0 E_n + P_n = \sigma, \tag{19.1}$$

where σ is the surface density of (free) charge on the conductor and all fields refer to the outer side; inside, they all vanish. If there are several conductors – separated by insulators, of course – the conditions (19.1) hold on the surface of each conductor.

The conditions (19.1) are just boundary conditions on the field in the insulating medium outside the conductors, and everything else depends on the properties of this medium. If it is a dielectric, we shall need information regarding $\mathbf{P}$ – in fact, a constitutive relation. We shall deal with dielectrics in the next chapter. If the insulator is non-polarizable, it is governed by the equations

$$\text{div}\, \epsilon_0 \mathbf{E} = q,$$

$$\mathbf{curl}\ \mathbf{E} = 0. \tag{19.2}$$

From $(19.2)_2$ we have $\mathbf{E} = -\,\mathbf{grad}\, V$, where the electric potential V includes an arbitrary additive constant (this is what the gauge transformation reduces to in electrostatics). In terms of the potential, equations (19.2) become

$$\Delta V = -q/\epsilon_0. \tag{19.3}$$

This is Poisson's equation. It is to be solved subject to the condition that V is constant on the surface of each conductor, which follows from $(19.1)_1$; each conductor is therefore an *equipotential*. Further conditions may be required, and we shall address them later on. Having found any particular solution, corresponding to a given distribution of q throughout the insulating medium outside the conductors, we can obtain other solutions by adding solutions of the *homogeneous* equation $\Delta V = 0$ – Laplace's equation – which also satisfy conditions of constancy of V on each conductor. For every solution, we obtain the surface charge density on any conductor from the boundary condition $(19.1)_2$ (with $\mathbf{P} = 0$)

$$\sigma = -\epsilon_0 \frac{\partial V}{\partial n}. \tag{19.4}$$

The total charge on a conductor is

$$Q = -\epsilon_0 \int \frac{\partial V}{\partial n}\, dS, \tag{19.5}$$

the integral being taken over the complete surface of the conductor.

The electric field contributes to the stress on the conductor. If the material outside the conductor is non-polarizable and belongs to one of the constitutive classes of §15 or §16, the electric part of the stress can be separated from the rest and is $-\frac{1}{2}\epsilon_0 E^2 I + \epsilon_0 \mathbf{E} \otimes \mathbf{E}$. The corresponding force per unit area is

$$T\mathbf{n} = -\mathbf{n}\tfrac{1}{2}\epsilon_0 E^2 + \epsilon_0 \mathbf{E} E_n = \mathbf{n}\tfrac{1}{2}\epsilon_0 E^2 = \tfrac{1}{2}\sigma \mathbf{E}, \tag{19.6}$$

where we have used the fact that $\mathbf{E}$ is normal to the conductor, and the relation $\epsilon_0 E_n = \sigma$. This force is normal to the conductor and directed outward – a tension, or a negative pressure. It tends to inflate the conductor at the expense of the surrounding insulator; this effect is called *electrostriction*. The result (19.6) is independent of the density of the material outside the conductor. It therefore holds in the limit in which this becomes a vacuum (a good vacuum will in any case be limited by the vapor pressure of the conductor); the expansion will then be halted by the elastic stress in the conductor that counteracts the tension (19.6). Note the factor of one half in the last member of (19.6). The force is sometimes said to be given by the charge density σ times the mean (between zero inside and $\mathbf{E}$ outside) electric field; far from explaining the $\frac{1}{2}$, this statement begs the question.

As a simple example, consider a spherical conductor of radius a that carries a total charge Q. If the charge density outside the conductor vanishes,

$$V = \frac{Q}{4\pi\epsilon_0 r}, \tag{19.7}$$

where r is the distance from the centre of the sphere, is a solution of the electrostatic problem. For the potential (19.7) is constant on the surface $r = a$, and the surface charge density $\sigma = -\epsilon_0 \partial V/\partial n$ has the uniform value $Q/(4\pi a^2)$, which adds up to Q over the whole surface. It remains to verify that $\Delta V = 0$ outside the sphere. Since $r = (x^2 + y^2 + z^2)^{1/2}$, we have (for $r > 0$)

$$\mathbf{grad}\,\frac{1}{r} = -\frac{1}{r^2}\,\mathbf{grad}\,r = -\frac{\mathbf{r}}{r^3}, \tag{19.8}$$

where $\mathbf{r} = (x, y, z)$. Furthermore (for $r > 0$),

$$\operatorname{div}\mathbf{grad}\,\frac{1}{r} = -\operatorname{div}\frac{\mathbf{r}}{r^3} = -\frac{1}{r^3}\operatorname{div}\mathbf{r} - \mathbf{r} \cdot \mathbf{grad}\,\frac{1}{r^3}$$

$$= -\frac{3}{r^3} + \mathbf{r} \cdot \frac{3}{r^4}\,\mathbf{grad}\,r = 0. \tag{19.9}$$

Any function U is called *harmonic* in a region if $\Delta U = 0$ in that region. Equation (19.9) shows that $1/r$ is harmonic in any region that does not include the point $r = 0$. We have thus demonstrated that (19.7) solves the electrostatic problem for a conductor whose outer boundary is a sphere. Since the force per unit area (19.6) is constant over the sphere for this solution, its resultant vanishes by symmetry. It does not matter whether the sphere is hollow or not. Equation (19.7) provides a solution even if the sphere contains an irregular wormhole in its interior. Whether it is the *only* solution is another matter. The following problem shows that it is not.

PROBLEM

Problem 19–1. If $\mathbf{E}_0$ is a constant vector, prove that

$$\mathbf{grad}(\mathbf{E}_0 \cdot \mathbf{r}) = \mathbf{E}_0. \tag{19.10}$$

Hence prove that

$$V = \frac{Q}{4\pi\epsilon_0 r} + \left(\frac{a^3}{r^3} - 1\right)(\mathbf{E}_0 \cdot \mathbf{r}) \tag{19.11}$$

is another solution for a spherical conductor of radius a with a total charge Q. This solution corresponds to a uniform electric field $\mathbf{E}_0$ at infinity. Prove that the resultant force $\int T\mathbf{n}\,dS$ is $Q\mathbf{E}_0$, equal to the force that the field $\mathbf{E}_0$ would exert on a particle carrying a charge Q.† This is *Coulomb's law*.

As a second example, consider a pair of infinite, parallel, conducting planes in vacuum. Let $y = \pm\frac{1}{2}d$ be the equations of the two planes. The linear function $V = (v/d)y$, with constant v (the potential difference between the plates), satisfies $\Delta V = 0$ and assumes the constant values $\pm\frac{1}{2}v$ on the conductors. Hence it is a solution. The surface charge densities on the conductors are $\pm\epsilon_0 v/d$, since $\partial V/\partial n$ is $\partial V/\partial y$ on one and $-\partial V/\partial y$ on the other. The total charge on each infinite plane is of course infinite. The electric field is uniform and normal to the planes; its direction is from the positively charged conductor to the negatively charged one. Finally, each conductor is attracted towards the other one by a force of $\frac{1}{2}\epsilon_0 v^2/d^2 = \frac{1}{2}\epsilon_0 E^2$ per unit area.

In order to discuss the properties of solutions of Poisson's equation (19.3) we shall make use of Green's identities, which follow from his divergence theorem (Gauss's formula). Let U and V be any sufficiently smooth functions. If we substitute the vector $\mathbf{a} = U\,\mathbf{grad}\,V$ in the

† Like the factor $\frac{1}{2}$ in (19.6), this result should not be regarded as obvious.

formula $\int \operatorname{div} \mathbf{a}\, d^3x = \int a_n\, dS$, we obtain

$$\int (U\Delta V + \operatorname{\mathbf{grad}} U \cdot \operatorname{\mathbf{grad}} V)\, d^3x = \int U\frac{\partial V}{\partial n}\, dS. \qquad (19.12)$$

This is Green's first identity. His second identity,

$$\int (V\Delta U - U\Delta V)\, d^3x = \int \left(V\frac{\partial U}{\partial n} - U\frac{\partial V}{\partial n}\right) dS, \qquad (19.13)$$

is a simple consequence of the first one.

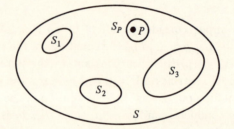

FIGURE 4

Consider now a series of conductors S_1, S_2,..., together with a distribution of charge density q in the medium outside the conductors. Let P be any point outside the conductors. We surround P by a very small sphere S_P, and by a large closed surface S, which includes S_P and S_1, S_2,...(Figure 4). Outside S there may be other conductors and charge distributions. Let V be a solution of Poisson's equation (19.3) inside S, and let $U = 1/r$, where r is the distance from P. We now apply Green's second identity to the region outside S_P and S_1, S_2,..., and inside S. On the left hand side we obtain $\int q/(\epsilon_0 r)\,d^3x$, since $\Delta V = -q/\epsilon_0$ and $\Delta r^{-1} = 0$. The right hand side is a sum of integrals over S_1, S_2,..., S_P and S in which $\partial/\partial n$ denotes differentiation out of the region, i.e. out of S and into S_1, S_2,... and S_P.

The contribution from the sphere S_P is $\int (V + r\partial V/\partial r)\,d\omega$, where r is the radius of S_P and $d\omega = dS_P/r^2$ the element of solid angle subtended at P by dS_P. When S_P collapses on P this gives $4\pi V(P)$, because V and $\mathbf{grad}\,V$ exist at P, as they do everywhere inside S for the solution V of $\Delta V = -q/\epsilon_0$.

The contribution from S_1 is obtained by putting $V = V_1$, the constant value of V on S_1, and noting that $\partial r^{-1}/\partial n = (\partial r^{-1}/\partial r)\cos\theta$, where θ is the angle between the radius from P and the inward normal to S_1. We have (Figure 5)

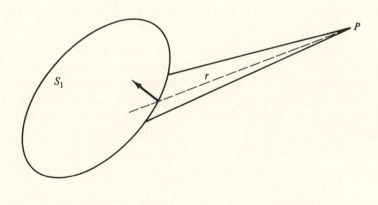

FIGURE 5

$$V_1 \int_{S_1} \frac{-\cos\theta}{r^2} \, dS = -V_1 \int_{S_1} d\omega = 0, \qquad (19.14)$$

since P lies outside S_1. The second term is $\int r^{-1} \partial V / \partial n \, dS$, in which $\partial / \partial n$ now denotes differentiation along the *outward* normal to S_1. According to (19.4), this is $\int \sigma / (\epsilon_0 r) \, dS$.

Collecting all the terms, we obtain

$$4\pi V(P) = \frac{1}{\epsilon_0} \int \frac{q \, d^3x}{r} + \frac{1}{\epsilon_0} \int \frac{\sigma \, dS}{r} - \int \left(V \frac{\partial}{\partial n} \frac{1}{r} - \frac{1}{r} \frac{\partial V}{\partial n} \right) \, dS, \quad (19.15)$$

where the second integral stands for a sum of integrals over the conductors, and the third is over S.

Let us now consider a *finite* system of conductors and charges. A solution V for a finite system is called *regular* if rV and $r^2 |\operatorname{grad} V|$ are bounded as r goes to infinity; of course, r is measured from a point P at finite distance from the system. (For the case of a spherical conductor with no charges outside it, (19.7) is a regular potential, but (19.11) is not.) If V is a regular potential, we can let S in (19.15) recede to infinity, and its contribution will vanish. Of course, we must ensure that the remaining integrals converge, but that is implied in the definition of a finite system. For a regular potential, then,

$$V = \frac{1}{4\pi\epsilon_0} \left(\int \frac{q \, d^3x}{r} + \int \frac{\sigma \, dS}{r} \right), \qquad (19.16)$$

where the first integral extends over all space and the second over all the conductors. Equation (19.16) is sometimes called the solution of Poisson's equation. It is no such thing, because σ stands for $-\epsilon_0 \partial V / \partial n$ and is in general unknown. But (19.16) has some important implications. In the first place, we may remove the conductors, so that only the charge distribution with density q remains. We then deduce that

$$V = \frac{1}{4\pi\epsilon_0} \int \frac{q \, d^3x}{r} \qquad (19.17)$$

is a (particular) regular solution of Poisson's equation (19.3). This can also be verified directly. Of course, if we now re-introduce one or several conductors, even in places where $q = 0$, it will cease to be a solution, because it does not satisfy the condition of constancy on the conductors. But (19.16) shows that the correct regular potential obeys the rule, '$4\pi\epsilon_0 V$ equals the sum of charges divided by their distances', even when conductors are present. Another implication of (19.16) is that it suggests a method of arriving at a solution of the electrostatic problem: we assume, on each one of the conductors, a surface charge distribution σ that depends on certain parameters; for a given set of values for the parameters, we treat (19.16) as a formula for V; we then try to adjust the parameters in such a way that V becomes constant on each conductor.

Unfortunately, there is no general method that leads with certainty to a solution of a given electrostatic problem. One is therefore forced to resort to guesses and tricks. But, having succeeded in obtaining a solution in one way or another, one may wonder whether there are others. Suppose, then, that two solutions of Poisson's equation exist which satisfy the same conditions: both are constant over each conductor; both yield the same total charges on those conductors whose charges are given, and the same potentials on those whose potentials are prescribed; finally, both have the same required value, or the same required gradient (electric field) at infinity. Now apply Green's first identity (19.12), with $V = U$, to the *difference* U of the two solutions. Since $\Delta U = 0$,

$$\int (\mathbf{grad}\, U)^2 \, d^3x = \int U \frac{\partial U}{\partial n} \, dS. \tag{19.18}$$

On the left hand side, the integral extends over all space outside the conductors. The right hand side is a sum of integrals over the conductors and the sphere at infinity. There is no contribution from the sphere at infinity, because either U or $\mathbf{grad}\, U$ vanish there. In each of the integrals over the conductors, U is the difference of two constants and is therefore itself a constant. Each of these integrals is therefore U/ϵ_0 times the difference between the total charges corresponding to the two solutions. On a conductor with prescribed potential, the first factor vanishes; on a conductor with prescribed charge, the second factor does. Hence the left hand side of (19.18) vanishes, and U is a constant everywhere. If the conditions of the problem prescribe the potential *somewhere*, the difference U will vanish there, and therefore everywhere. Otherwise any solution is determined to within an additive constant, and we conclude, again, that any two solutions are essentially the same.

For a single conductor in empty space, let V be the regular solution that corresponds to a constant potential V_1 and a charge Q_1 on the conductor. If λ is any constant then, by the uniqueness theorem we have just proved, λV is *the* regular solution that corresponds to the constant potential λV_1 and the charge λQ_1 on the conductor. The ratio

$$C = \frac{Q}{V} = -\frac{\epsilon_0}{V} \int \frac{\partial V}{\partial n} \, dS \qquad (19.19)$$

is therefore a property of the conductor which is purely geometric. It is called the *capacity* of the conductor. Its SI unit is the *farad*, defined by 1 farad = 1 coulomb/volt. For a conductor of given shape, it is proportional to the size, as is evident from (19.19). It is easy to prove that $C > 0$. According to (19.7), $C = 4\pi\epsilon_0 a$ for a spherical conductor of radius a.

A *capacitor*, or *condenser*, is an instrument for storing charge. It consists of two conductors of arbitrary shapes called the *plates* of the capacitor. The positive plate carries a positive charge Q_+ and the negative plate a charge $Q_- = -Q_+$, so that the net charge on the capacitor as a whole vanishes. The capacity of a capacitor is defined by

$$C = \frac{Q_+}{V_+ - V_-}. \qquad (19.20)$$

Again, this is a purely geometrical constant depending on the position and shape of the two plates.

The arrangement we have considered above of two infinite, parallel, plane conductors constitutes a capacitor. Since the charges on the plates are infinite, we define a capacity per unit area, equal to the charge per unit area divided by the potential difference, or *voltage*. We have seen that the charge per unit area is $\pm\epsilon_0 v/d$, where v is the voltage. Hence the capacity per unit area is ϵ_0/d. If the parallel plates are finite rather than infinite, there will be edge effects, but we expect these to be small so long as the plate diameter is very large compared to the distance between them. For a parallel-plate capacitor in which the area of a plate is S ($S \gg d^2$), the capacity is

$$C = \frac{\epsilon_0 S}{d}. \qquad (19.21)$$

Equation (19.18) can also be used to prove that a function which is harmonic in a region is uniquely determined by its values (constant or not) on the complete boundary of the region (Dirichlet's problem); alternatively, it is determined up to an additive constant by the values of its normal derivative on the complete boundary (Neumann's problem).

The potential problem in its various forms does not have more than one solution. But does it have one? This question of *existence* is far more difficult than that of uniqueness, and has occupied the best mathematicians for a century. It has been proved that there is a solution if the boundary is sufficiently regular and the prescribed values on it sufficiently smooth. A precise statement of these conditions lies outside the scope of this book. We shall simply assume that they are satisfied in every problem we shall discuss. After all, the mathematical examples we usually treat are stated in terms of analytic functions (the smoothest one can think of) for the equations of the surfaces and for the prescribed values of V or $\partial V/\partial n$. The experimenter, on the other hand, either models his arrangement in mathematical terms in order to apply the theory, and then solves a mathematical example; or he just measures capacities and potential differences, in which case he is not concerned with their existence.

As an example that uses the uniqueness of Neumann's problem, consider the following magnetostatic problem, which arises in discussions of the magnetic confinement of a plasma: within a simply-connected region (like the interior of a sphere) it is desired to set up a given, static, magnetic field $\mathbf{B}$. Can this be achieved by suitable surface currents on the boundary S of the region (like the seam on a tennis ball)? We first note that the given field inside S has a given normal component B_n on S, and this must be continuous. If there are to be no currents outside S, the external field must satisfy $\operatorname{div}\mathbf{B}_e = 0$ and $\operatorname{\mathbf{curl}}\mathbf{B}_e = 0$. It follows that $\mathbf{B}_e = \operatorname{\mathbf{grad}}\Omega$ and $\Delta\Omega = 0$, with $\partial\Omega/\partial n = B_n$ given on S. Since this Neumann's problem has a unique regular solution, $\mathbf{B}_e$ is determined, and so is the surface current $\mathbf{K} = \mathbf{n} \times (\mathbf{B}_e - \mathbf{B})/\mu_0$ on S.

PROBLEMS

Problem 19–2. Show that, if a function $f(\mathbf{x})$ has a minimum at P, its Laplacian at P is positive. Deduce that, in a system of charged conductors with no volume charges between them, the potential attains its minimum or its maximum on one of the conductors or at infinity.

Problem 19-3. Show that, in a system of charged conductors with zero total net charge, and with no volume charges, at least one conductor is everywhere charged positively, and one everywhere negatively. A capacitor is a specially simple case to which this theorem applies.

Problem 19-4. A spherical soap bubble carries a charge Q. Its surface tension provides an inward force of $2\alpha/r$ per unit area. Determine its equilibrium radius.

Problem 19-5. Three concentric hollow conducting spheres have radii a, b and c. The inner and outer are connected by a fine wire and form one plate of a capacitor. The middle sphere is the other plate. Show that the capacity is $4\pi\epsilon_0 b^2(c-a)/[(b-a)(c-b)]$.

Problem 19-6. A series of conductors carrying charges Q_1, Q_2,... have potentials V_1, V_2,... . Let Q'_1, Q'_2,... be a second set of charges on the same conductors, corresponding to potentials V'_1, V'_2,... . Prove Green's reciprocal theorem:

$$\sum Q_a V'_a = \sum Q'_a V_a.$$

Show that the result still holds if some, or all, of the conductors are replaced by point charges, and the potential V_i at a point charge Q_i is defined as the potential due to all charges except Q_i.

Problem 19-7. Apply Green's reciprocal theorem to the case when all the charges are zero, except Q_1 and Q'_2, the latter being a point charge. Deduce that the potential of an uncharged conductor under the influence of a unit charge at a point P is the same as the potential at P due to a unit charge placed on the conductor. Thus a unit charge at a distance d from the centre of an uncharged conducting sphere raises the latter to potential $(4\pi\epsilon_0 d)^{-1}$.

20. Multipole moments

If the charge in the particular solution (19.7) is concentrated in particles, the formula becomes

$$V = \frac{1}{4\pi\epsilon_0} \sum \frac{e_a}{r_a}. \tag{20.1}$$

We recall that $\mathbf{r}_a$ is the vector from the point P, to which V refers, to the charge e_a. We choose the origin at some point O which is at a finite distance from the particles and denote by $\mathbf{R}$ the position of P relative to O. Then

$$V(\mathbf{R}) = \frac{1}{4\pi\epsilon_0} \sum \frac{e_a}{|\mathbf{R} - \mathbf{r}_a|}. \tag{20.2}$$

At great distance from the system ($R \gg r_a$), we expand each $|\mathbf{R} - \mathbf{r}_a|^{-1}$ according to the formula

$$f(\mathbf{R} - \mathbf{r}) = f(\mathbf{R}) - \mathbf{r} \cdot f_{\mathbf{R}}(\mathbf{R}) + \tfrac{1}{2} r_i r_j f_{R_i R_j}(\mathbf{R}) + \dots . \tag{20.3}$$

The first term is

$$V^{(0)} = \frac{1}{4\pi\epsilon_0} \frac{\sum e_a}{R}. \tag{20.4}$$

It is the potential at $\mathbf{R}$ of a single charge of amount $\sum e_a$ located at the origin; $\sum e_a$ is called the *monopole moment* of the system. The corresponding field is

$$\mathbf{E}^{(0)} = -\operatorname{grad} V^{(0)} = \frac{1}{4\pi\epsilon_0} \frac{\sum e_a}{R^2} \mathbf{n}, \tag{20.5}$$

where $\mathbf{n}$ is a unit vector along $\mathbf{R}$.

The second term is

$$V^{(1)} = -\sum e_a \mathbf{r}_a \cdot \frac{\partial}{\partial \mathbf{R}} \frac{1}{R} = \frac{\mathbf{d} \cdot \mathbf{R}}{R^3} = \frac{\mathbf{d} \cdot \mathbf{n}}{R^2}, \tag{20.6}$$

where

$$\mathbf{d} = \sum e_a \mathbf{r}_a \tag{20.7}$$

is the *dipole moment* of the system.

PROBLEM

Problem 20–1. Show that the dipole moment is independent of the choice of origin if the monopole moment $\sum e_a$ vanishes.

In particular, for two charges with $e_1 = e$ and $e_2 = -e$,

$$\mathbf{d} = e_1 \mathbf{r}_1 + e_2 \mathbf{r}_2 = e(\mathbf{r}_1 - \mathbf{r}_2) = e\mathbf{r}, \tag{20.8}$$

where $\mathbf{r}$ is the position of $e_1 = e$ relative to e_2. The dipole field corresponding to the dipole potential (20.6) is

$$\mathbf{E}^{(1)} = -\operatorname{grad} V^{(1)} = \frac{3(\mathbf{d} \cdot \mathbf{n})\mathbf{n} - \mathbf{d}}{R^3}. \tag{20.9}$$

The third term in the expansion of (20.2) according to (20.3) is

$$V^{(2)} = \tfrac{1}{2} \sum er_i r_j \frac{\partial^2}{\partial R_i \partial R_j} \frac{1}{R}, \tag{20.10}$$

where $\sum$ denotes summation over all the charges, as before; for each charge, there is a summation over each of the repeated indices i and j.

PROBLEM

Problem 20–2. Show that

$$V^{(2)} = \frac{D_{ij} n_i n_j}{2R^3}, \tag{20.11}$$

where

$$D_{ij} = \sum e(3r_i r_j - r^2 \delta_{ij}). \tag{20.12}$$

The symmetric tensor D of (20.12) is called the *quadrupole moment* of the system. It is evidently traceless. Hence it has five independent components.

PROBLEM

Problem 20–3. Prove that D does not depend on the choice of origin if $\sum e$ and $\mathbf{d}$ both vanish.

The expansion of the potential (20.2) according to (20.3) becomes rather unwieldy beyond the quadrupole term. It is easier to proceed to higher multipole moments by use of the formula

$$\frac{1}{|\mathbf{R} - \mathbf{r}|} = \frac{1}{\sqrt{R^2 + r^2 - 2rR\cos\chi}} = \sum_{n=0}^{\infty} \frac{r^n}{R^{n+1}} P_n(\cos\chi), \tag{20.13}$$

where χ is the angle between $\mathbf{R}$ and $\mathbf{r}$, and $P_n(x)$ is the Legendre polynomial of degree n. We shall return to these matters in §21.

We have based the discussion of multipole moments on the formula (20.1). Since the expansion was performed on $|\mathbf{R} - \mathbf{r}|^{-1}$, it is equally possible to carry it out for continuously distributed charges, starting from (19.17). The only difference is that the moments become integrals over the charge density, instead of sums over the discreet charges.

PROBLEM

Problem 20–4. An electric dipole is at the origin, and its direction is that of the z axis, so that the potential (20.6) is proportional to $(\cos\theta)/r^2$. An element of a field line has radial and transverse projections dr and $r\,d\theta$ such that $dr : r\,d\theta = E_r : E_\theta$. Show that the field lines are given by $(\sin^2\theta)/r = \text{const.}$

21. Methods for solving electrostatic problems

We have already noted that there exists no general method for solving every given electrostatic problem. There are, however, several methods which have been found successful in dealing with certain classes of problems. A new problem may, with luck, turn out to belong to one of these classes; or it may yield to a combination of these methods.

The first method is to expand the potential V as a series of harmonic functions with coefficients that must be determined from the boundary conditions – the prescribed potentials or charges on conductors and the behaviour at infinity.

In spherical coordinates (r, θ, ϕ) the simplest harmonic functions are

$$r^n P_n(\cos\theta) \qquad \text{and} \qquad r^{-(n+1)} P_n(\cos\theta),$$

where n is an integer. For $n \le 1$ these are

$$1, \qquad r\cos\theta = z, \qquad 1/r \qquad \text{and} \qquad \cos\theta/r^2.$$

The functions $r^n P_n(\cos\theta)$ are finite at $r = 0$, but not at infinity; the opposite is true of the functions $r^{-(n+1)} P_n(\cos\theta)$. Both types have rotational symmetry around the axis $\theta = 0$, and their use is therefore confined to problems that have this symmetry. Otherwise, each Legendre polynomial $P_n(\cos\theta)$ must be replaced by a linear combination of the $2n + 1$ spherical harmonics $Y_{nm}(\theta, \phi)$, with $-n \le m \le n$.

In cylindrical coordinates (r, θ, z) the standard harmonic functions are

$$J_m(nr)\cos m\theta e^{\pm nz} \qquad \text{and} \qquad J_m(nr)\sin m\theta e^{\pm nz},$$

where $J_m(nr)$ is the Bessel function of order m and argument nr.

In two-dimensional polar coordinates (r, θ) the corresponding functions are

$$r^n \cos n\theta \qquad \text{and} \qquad r^n \sin n\theta.$$

As an example, consider the problem of a spherical conductor in a uniform field $\mathbf{E}_0$ parallel to the z axis. We already know that the (unique) solution is given by (19.11), but now we wish to derive it. At infinity, V must tend to $-\mathbf{E}_0 \cdot \mathbf{r} = -E_0 r \cos \theta$; hence the difference $V - (-E_0 r \cos \theta)$ must be finite. We therefore try

$$V = -E_0 r \cos \theta + \sum_{n=0}^{\infty} \frac{A_n}{r^{n+1}} P_n(\cos \theta). \tag{21.1}$$

This satisfies the equation $\Delta V = 0$ and the boundary condition $V \to -\mathbf{E}_0 \cdot \mathbf{r}$. It remains to impose the conditions that V be a constant V_0 on sphere $r = a$ and that the total charge on the conductor be a given Q. The first gives

$$0 = (\frac{A_0}{a} - V_0) P_0(\cos \theta) + (\frac{A_1}{a^2} - E_0 a) P_1(\cos \theta) + \sum_{n=2}^{\infty} \frac{A_n}{a^{n+1}} P_n(\cos \theta)$$

for all θ. Now the Legendre polynomials are linearly independent. Hence

$$A_0 = a V_0, \qquad A_1 = E_0 a^3,$$

and $A_2 = A_3 = \ldots = 0$. The charge density on $r = a$ is

$$\sigma = -\epsilon_0 \frac{\partial V}{\partial r} = \epsilon_0 E_0 \cos \theta + \sum_{n=0}^{\infty} (n+1) \frac{\epsilon_0 A_n}{a^{n+2}} P_n(\cos \theta).$$

The total charge is $Q = \int \sigma \, dS = a^2 \int \sigma \, d\omega$, but only the term with $P_0(\cos \theta)$ can survive this integration. Hence $Q = 4\pi\epsilon_0 A_0$. The solution is therefore

$$V = -E_0 r \cos \theta + \frac{Q}{4\pi\epsilon_0 r} + \frac{E_0 a^3}{r^2} \cos \theta;$$

this is (19.11).

The second method is the method of images.† We present it by an example. If an electron is ejected from a metal (perhaps as a result

---

† It is due to Kelvin.

of 'thermal agitation'), what is the force with which it is being pulled
back? Consider, then, a particle with charge $-e$ at a distance x from a
conducting half-space. The conductor is left with the charge $+e$, which
must be distributed over its plane surface. We wish to determine the
field outside the conductor. If we replace the charge on the conductor
by a particle of charge e placed at the *image* point – with respect to the
surface of the conductor – of the electron, the potential becomes

$$V = \frac{1}{4\pi\epsilon_0}(\frac{e}{r'} - \frac{e}{r}); \qquad (21.2)$$

here r is the distance from the electron and r' is the distance from the
image point. Outside the conductor this potential satisfies Laplace's
equation. It is also constant on the surface of the conductor. Hence it
is the solution. The field of the image has the magnitude $e/(16\pi\epsilon_0 x^2)$
at the position of the electron. The force is therefore $e^2/(16\pi\epsilon_0 x^2)$.

PROBLEM

Problem 21–1. Calculate the surface charge density σ on the conductor. Prove
that $\int \sigma \, dS = e$.

The method of images is evidently capable of generalization. The problem of several particles in front of a plane is as easy as the one we have
considered: for each particle, we introduce an image. For a finite body
in front of a plane, we introduce its finite image, and so on. But the
method is not confined to images in a plane.

Let e be a point charge outside a spherical conductor of radius a, at a
distance x ($> a$) from the centre. We place an image charge $-e'$ inside
the sphere, at a distance x' from the centre along the line joining the
centre to e, and seek to determine e' and x' in such a way that

$$V = \frac{1}{4\pi\epsilon_0}(\frac{e}{r} - \frac{e'}{r'}) \qquad (21.3)$$

will vanish on the sphere (r and r' are the distances from e and e').

PROBLEM

Problem 21-2. Prove that e' and x' are given by

$$e' = \frac{a}{x}e, \qquad x' = \frac{a^2}{x}. \tag{21.4}$$

With e' and x' given by (21.4), the potential (21.3) vanishes on the sphere. By Gauss's theorem, it is the solution corresponding to the case in which the spherical conductor carries a charge $-e'$. If the conductor is to have a total charge Q, we add a point charge $Q + e'$ at the centre; this will not disturb the constancy of V on the sphere.

The third method uses functions of a complex variable for potential problems in two dimensions. Laplace's equation is then

$$\Delta V = V_{xx} + V_{yy} = 0. \tag{21.5}$$

We put $z = x + iy$, where $i = \sqrt{-1}$, and consider the function

$$w = f(z) = V(x, y) + iU(x, y); \tag{21.6}$$

we have denoted the real part of w by V and the imaginary part by U. Differentiation of $w = f(x + iy)$ gives $w_{xx} + w_{yy} = 0$. Separating real and imaginary parts, we have

$$\Delta V = 0, \qquad \Delta U = 0. \tag{21.7}$$

Hence V and U are harmonic. Furthermore,

$$w_y = if'(z) = iw_x. \tag{21.8}$$

Equating the real and imaginary parts, we obtain Cauchy and Riemann's relations

$$V_x = U_y, \qquad U_x = -V_y. \tag{21.9}$$

PROBLEM

Problem 21-3. Prove that $V(x, y) = $ const. and $U(x, y) = $ const. are two families of curves which cut orthogonally.

If we regard the real part V of w as the electric potential, the lines $V =$ const. become the equipotentials; therefore the lines $U =$ const. are the electric field lines. With this interpretation, $w(z) = V + iU$ is called the complex potential.†

PROBLEMS

Problem 21–4. Prove that the magnitude of the electric field is

$$E = |w'(z)|. \tag{21.10}$$

Problem 21–5. Let c be a closed curve in the (x, y) plane, and let $\delta(U)$ be the change in U upon traversing c in a counterclockwise direction. Prove that $-\delta(U)/\epsilon_0$ is the charge – per unit length in the z direction – inside c.

As a simple example, consider the complex potential

$$w = -\frac{\lambda}{2\pi\epsilon_0} \ln z. \tag{21.11}$$

If we use polar coordinates (r, θ), then $z = re^{i\theta}$ and $\ln z = \ln r + i\theta$. The real part of (21.11) becomes

$$V = -\frac{\lambda}{2\pi\epsilon_0} \ln r, \tag{21.12}$$

which is the potential of a line, perpendicular to the (x, y) plane through the origin, that carries a charge λ per unit length. The imaginary part of w is $\lambda\theta/(2\pi\epsilon_0)$. Hence $\theta =$ const. gives the electric field lines. If we substitute $z - z_0$ for z in (21.11) we get the complex potential for a line through $z_0 = x_0 + iy_0$ instead of through the origin.

A more interesting example is provided by the complex potential defined (implicitly) through

$$z = 1 + iw + e^{iw}. \tag{21.13}$$

† Some authors identify the potential with the imaginary part of w; others write $w = V - iU$.

Separating real and imaginary parts, we have

$$x = 1 - U + e^{-U} \cos V,$$

$$y = V + e^{-U} \sin V. \tag{21.14}$$

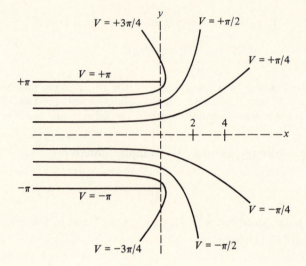

FIGURE 6

Consider now the two equipotentials $V = \pm\pi$. Along these,

$$x = 1 - U - e^{-U}, \qquad y = \pm\pi.$$

Since $e^{-U} > 1 - U$ for any U, these equipotentials are the straight half-lines $x < 0$, $y = \pm\pi$. We identify them with the two plates of a capacitor. If U is eliminated between the two members of (21.14), the result,

$$x = 1 + \ln(y - V) - \ln \sin V + (y - V) \cot V, \tag{21.15}$$

gives an equipotential $x = x(y)$ for each value of V between $-\pi$ and π. Some of these are shown in Figure 6. We see how the equipotentials fan out near the edge of a parallel-plate capacitor. If the plates are at the potentials $\pm v/2$, instead of $\pm\pi$, we must replace w by $wv/(2\pi)$ in (21.13); if they are at $y = \pm d/2$, instead of $\pm\pi$, we must replace z by $zd/(2\pi)$.

Obviously, the method of complex potentials (also called the method of conformal mapping) can be combined with the method of images to solve potential problems in two dimensions. As an example, we consider the circle theorem.† Let $f(z)$ be a complex potential with sources lying outside the circle $|z| = R$. If we now introduce a cylindrical conductor $|z| = R$ with its axis through the origin, what is the new complex potential? Consider the complex function

$$w = f(z) - \bar{f}(R^2/z). \tag{21.16}$$

Since $f(z)$ has no singularities inside the circle, $\bar{f}(R^2/z)$ has none outside, and its subtraction will not alter the distribution of sources outside the circle. On $|z| = R$, $\bar{f}(R^2/z) = \bar{f}(\bar{z})$ is the complex conjugate of $f(z)$, so that w is purely imaginary. In particular, the real part of w is constant on the circle. Hence (21.16) provides the solution. The singularities of $\bar{f}(R^2/z)$, which lie inside the circle, are the images required for the constancy of the electric potential on the cylindrical conductor.

PROBLEMS

Problem 21–6. Calculate the potential of a uniformly charged circular wire of radius a. Use the expansion $V = \sum(A_n r^n + B_n/r^{n+1})P_n(\cos\theta)$ and determine the coefficients A_n and B_n from the potential on the axis, where $r = z$ and $P_n(\cos\theta) = 1$. The latter can be obtained from (19.7) and expanded in powers of z.

Problem 21–7. A point charge e is placed inside a spherical cavity of radius a cut out of a conducting block of metal at zero potential. If the charge is at a distance d from the centre of the cavity, show that the force on it is $e^2 ad/[4\pi\epsilon_0(a^2 - d^2)^2]$.

Problem 21–8. Find the potential of a line charge lying outside and parallel to a cylinder.

† Milne-Thomson (1940)

CHAPTER VII

Dielectrics

22. Linear dielectrics

If the insulating medium in an electrostatic problem is a dielectric, we must consider the polarization $\mathbf{P}$. For concreteness, we shall do so within the framework of the constitutive classes we have introduced for fluids and solids in §15–16. We recall that, among these materials, the insulators are incapable of undergoing irreversible processes. If they are also non-magnetizable, $\mathcal{M} = -\rho\varphi_{\mathbf{B}} = 0$, and the free energy is independent of the magnetic field $\mathbf{B}$. So is

$$\mathbf{P} = -\rho\varphi_{\mathcal{E}}. \tag{22.1}$$

An immediate consequence of this relation is that $\partial P_i/\partial \mathcal{E}_j = \partial P_j/\partial \mathcal{E}_i$.

A *linear* dielectric is defined as one in which $\mathbf{P}$ and $\mathcal{E}$ are linearly related:

$$\mathbf{P} = \epsilon_0 \chi \mathcal{E}, \tag{22.2}$$

where χ, the *dielectric susceptibility* or the *polarizability*, is a matrix with elements that may depend on the temperature and the deformation (or the density). From the remark following (22.1), χ is symmetric. Integration of (22.1) gives, for a linear dielectric,

$$\rho\varphi = -\tfrac{1}{2}\mathbf{P} \cdot \mathcal{E} + \rho\psi, \tag{22.3}$$

where ψ is independent of $\mathcal{E}$. It is the specific free energy in the absence of an electromagnetic field, and it can, in principle, be determined from measurements carried out with $\mathbf{E} = \mathbf{B} = 0$. We may, of course, substitute for ψ any of the standard forms from classical thermodynamics, such as the specific free energy of a perfect gas or a solution. The product $\mathbf{P} \cdot \boldsymbol{\mathcal{E}} = \epsilon_0 \boldsymbol{\mathcal{E}}^T \chi \boldsymbol{\mathcal{E}}$ is a quadratic form in the components of $\boldsymbol{\mathcal{E}}$.

In an isotropic material the relation (22.2) must be unaffected by rotations. It can be shown that the matrix χ then becomes a multiple χ of the identity,† and (22.3) reduces to

$$\rho\varphi = -\tfrac{1}{2}\epsilon_0\chi\mathcal{E}^2 + \rho\psi. \tag{22.4}$$

For a linear, isotropic dielectric at rest, the displacement is

$$\mathbf{D} = \epsilon_0\mathbf{E} + \mathbf{P} = \epsilon_0(1 + \chi)\mathbf{E} = \epsilon\mathbf{E}, \tag{22.5}$$

where

$$\epsilon = (1 + \chi)\epsilon_0 \tag{22.6}$$

is called the *permittivity* or the *dielectric constant*. The dielectric susceptibility χ of isotropic dielectrics is always positive. It ranges from less than 10^{-3} (in gases at room temperature and atmospheric pressure) to more than 2000. Water has a χ of about 80. The permittivity is of course positive, too; in fact, $\epsilon > \epsilon_0$. The electrostatics of such dielectrics is governed by the equations

$$\mathbf{curl}\ \mathbf{E} = 0,$$

$$\operatorname{div} \mathbf{D} = \operatorname{div} \epsilon\mathbf{E} = q,$$

$$\mathbf{n} \times [\![\mathbf{E}]\!] = 0, \qquad \mathbf{n} \cdot [\![\mathbf{D}]\!] = \sigma. \tag{22.7}$$

If ϵ is uniform, these lead to Poisson's equation $\Delta V = -q/\epsilon$; in the absence of free charges (which would have no way of getting into an insulating dielectric) they lead to Laplace's equation $\Delta V = 0$. At a boundary between two dielectrics, both $\mathbf{n} \times \mathbf{E}$ and D_n are continuous. In terms of the potential, the conditions are:

$$V \quad \text{and} \quad \epsilon\frac{\partial V}{\partial n} \quad \text{are continuous.} \tag{22.8}$$

---

† In a fluid $\varphi(\rho, \vartheta, \boldsymbol{\mathcal{E}})$ must be unaffected by rotations. It follows, irrespective of any considerations of isotropy, that φ can only depend on $\mathcal{E}^2$.

As an example, we consider a dielectric sphere of radius a in an external uniform field $\mathbf{E}_0$ which is parallel to the z direction. Let ϵ_i be the permittivity inside the sphere, and ϵ_e the permittivity of the medium outside it. For the potential V_e outside the sphere we assume the form (21.1). Inside the sphere we set $V_i = \sum B_n r^n P_n(\cos\theta)$, which is finite at $r = 0$. The conditions (22.8) lead to

$$\mathbf{E}_e = \mathbf{E}_0 + \frac{a^3}{r^3}\frac{\epsilon_i - \epsilon_e}{2\epsilon_e + \epsilon_i}[3(\mathbf{n}\cdot\mathbf{E}_0)\mathbf{n} - \mathbf{E}_0],$$

$$\mathbf{E}_i = \frac{3\epsilon_e}{2\epsilon_e + \epsilon_i}\mathbf{E}_0. \tag{22.9}$$

Somewhat unexpectedly, the field inside the sphere turns out to be uniform.

At a conductor–dielectric boundary there is still a free surface charge density, and the relation (19.4) is replaced by

$$\sigma = D_n = -\epsilon\frac{\partial V}{\partial n}. \tag{22.10}$$

As a simple example, we consider a spherical conductor of radius a that carries a charge Q and is embedded in a dielectric medium with permittivity ϵ. The potential is then given by (19.7), with ϵ_0 replaced by ϵ.

Similarly, if we substitute ϵ for ϵ_0 in the solution for the infinite, parallel-plate, capacitor, we obtain the solution for the case in which the medium between the plates is a dielectric with permittivity ϵ. The capacity per unit area is ϵ/d, larger than the ϵ_0/d for a vacuum.

PROBLEMS

Problem 22–1. A parallel-plate capacitor is filled with two dielectrics, arranged in alternating layers that are all parallel to the plates. The layers of the first dielectric, with permittivity ϵ_a, occupy a fraction a (< 1) of the volume between the plates; the layers of the second dielectric, with permittivity ϵ_b, a fraction $b = 1 - a$. Prove that the capacity is $\epsilon S/d$, where $\epsilon^{-1} = a\epsilon_a^{-1} + b\epsilon_b^{-1}$.

Problem 22–2. The same as Problem 22–1, but with the two dielectrics arranged in prisms (or columns) perpendicular to the plates. Prove that the capacity is $\epsilon S/d$, with $\epsilon = a\epsilon_a + b\epsilon_b$.

Problem 22–3. A hollow cylinder, with inner radius a and outer radius b, consists of dielectric material with permittivity ϵ. The inside and outside surfaces are coated with conducting material. Neglecting edge effects, prove that if the length of the cylinder is L, its capacity is

$$C = \frac{2\pi\epsilon L}{\ln(b/a)}. \tag{22.11}$$

Let us now consider an uncharged dielectric medium of permittivity ϵ (a positive, but otherwise arbitrary, function of position) in which conductors carrying given charges have been placed. According to the first principle of electromagnetism (the first pair of Maxwell's equations), the displacement $\mathbf{D}$ must then be such that $\operatorname{div}\mathbf{D} = 0$ in the medium and the integrals $Q_a = \oint D_n \, dS_a$ over the conductors $a = 1, 2, \ldots$ have fixed values. We shall *not* assume that $\operatorname{\mathbf{curl}}\mathbf{E} = \operatorname{\mathbf{curl}}\mathbf{D}/\epsilon = 0$, i.e. $\mathbf{D} = -\epsilon\operatorname{\mathbf{grad}}V$, or that the charge density D_n on the conductors is distributed in such a way that V is constant over each conductor. In fact, we shall ignore the second and third principles (Maxwell's second pair and the aether relations). Rather, we shall consider the *functional*

$$W[\mathbf{D}] = \int \frac{D^2}{2\epsilon} \, d^3x, \tag{22.12}$$

the integral being taken throughout the medium; it will exist if the system of conductors is finite in the sense of §19. For any vector field $\mathbf{D}$ such that $\operatorname{div}\mathbf{D} = 0$ and $\oint D_n \, dS_a = Q_a$, $W[\mathbf{D}]$ is positive (since $\epsilon > 0$). It can be made arbitrarily large by choosing a sufficiently tangled $\mathbf{D}$ (we recall that $\operatorname{\mathbf{curl}}\mathbf{D}$ is now arbitrary). But W must have a minimum; in fact, a *positive* minimum (unless the charges Q_a on the conductors all vanish). In order to find the displacement that renders $W[\mathbf{D}]$ a minimum, we introduce Lagrangian multipliers $V(\mathbf{x})$ and λ_a and take the *unrestricted* variation of

$$W'[\mathbf{D}] = W[\mathbf{D}] - \int V \operatorname{div}\mathbf{D}\, d^3x - \sum \lambda_a \oint D_n \, dS_a \tag{22.13}$$

with respect to $\mathbf{D}$. The result, after an integration by parts (remembering that on each conductor δD_n is the normal component of $\mathbf{D}$ in the direction pointing outward, and hence *into* the dielectric medium), is

$$\delta W' = \int \delta \mathbf{D} \cdot (\mathbf{D}/\epsilon + \mathbf{grad}\,V)\,d^3x - \sum \oint \delta D_n(\lambda_a - V)\,dS_a. \quad (22.14)$$

If $\delta W'$ is to vanish for all $\delta \mathbf{D}$ throughout the medium, and for all δD_n on the conductors, then $\mathbf{D} = -\epsilon\,\mathbf{grad}\,V$, where V is constant over each conductor. These are precisely the requirements we have left out. This proves Thomson's theorem: the charges on the conductors distribute themselves in such a way as to render $W[\mathbf{D}]$ a minimum.

Thomson's theorem is often regarded as a statement about minimal energy, or minimal free energy. We have been more cautious. In the absence of motion and magnetic fields, substitution of the relations (22.4)–(22.6) for a linear dielectric into (14.10) gives for the energy density the expression

$$\rho(\psi + \vartheta\eta) + D^2/(2\epsilon). \quad (22.15)$$

Whereas ψ is the specific free energy of the dielectric in the *absence* of an electromagnetic field, the specific entropy $\eta = -\varphi_\vartheta$ *does* depend on the field: according to (22.4), φ_ϑ and ψ_ϑ can be equal only when the permittivity happens to be independent of the temperature. Thus, in (22.15), the combination $\psi + \vartheta\eta$ is *not* the specific energy in the absence of a field. Hence W, the object of Thomson's theorem, cannot generally be identified as the electrostatic energy (or free energy) of a dielectric medium.

What, then, is the content of Thomson's theorem? The answer follows from the statement of the theorem: it is a variational formulation of the electrostatic problem for a linear dielectric medium. This may have interesting implications. For example, the theorem suggests a method for solving electrostatic problems: among a class of displacements $\mathbf{D}$ depending on some parameters (such as the coefficients in an expansion in terms of known elementary solutions), the 'best' $\mathbf{D}$ is obtained by choosing those values of the parameters that render $W[\mathbf{D}]$ a minimum. This $\mathbf{D}$ may be the exact solution if we have been lucky, or clever, enough to have chosen a class that includes the exact solution. Another interesting conclusion is that the displacement field will tend to be larger in absolute value (the lines of force of the vector field $\mathbf{D}$ more concentrated) in regions of high permittivity, because this minimizes the integral of D^2/ϵ.

PROBLEMS

Problem 22–4. Extend the statements in Problems 19–2, 19–3 and 19–6 to charged conductors in a linear dielectric medium.

Problem 22–5. A point charge e is placed at a point P outside a semi-infinite medium of uniform permittivity ϵ. Show that in the vacuum the potential is the same as that due to a charge e at P and e' at the image point P', and in the dielectric it is the same as that due to a charge e'' at P, where

$$ e' = -\frac{\epsilon - \epsilon_0}{\epsilon + \epsilon_0}e, \qquad e'' = \frac{2\epsilon}{\epsilon + \epsilon_0}e. $$

23. Forces on linear dielectrics

In order to obtain the forces on fluid dielectrics and on conductors in contact with fluid dielectrics we use (22.4) to calculate the pressure, and obtain

$$ p = \rho^2 \varphi_\rho = \rho^2 \psi_\rho + \tfrac{1}{2}\epsilon_0(\chi - \rho\chi_\rho)\mathcal{E}^2, \tag{23.1} $$

where the first term, $\rho^2\psi_\rho = \pi$ (say), is the pressure in the absence of $\mathcal{E}$. If we substitute this in the formula (15.15) for the stress tensor of a fluid, we obtain in the electrostatic case ($\mathcal{M} = \mathbf{B} = \dot{\mathbf{x}} = 0$)

$$ T = -(\pi + \tfrac{1}{2}\mathbf{E} \cdot \mathbf{D} - \tfrac{1}{2}\epsilon_0\rho\chi_\rho E^2)I + \mathbf{E} \otimes \mathbf{D}. \tag{23.2} $$

As an example, we consider a dielectric with constant χ. At a static boundary, the components of the contact force,

$$ T\mathbf{n} = -(\pi + \tfrac{1}{2}\mathbf{E} \cdot \mathbf{D})\mathbf{n} + \mathbf{E}D_n, \tag{23.3} $$

must be continuous (cf. (12.3)). The tangential component of $T\mathbf{n}$ is $\mathbf{n} \times T\mathbf{n} = \mathbf{n} \times \mathbf{E}D_n$. The factor $\mathbf{n} \times \mathbf{E}$ is continuous at any stationary boundary. At a dielectric–dielectric boundary D_n, too, is continuous. At a conductor–dielectric boundary D_n is discontinuous, but $\mathbf{n} \times \mathbf{E} = 0$. Hence we need only worry about the normal component,

$$ \mathbf{n} \cdot T\mathbf{n} = -\pi - \tfrac{1}{2}\mathbf{E} \cdot \mathbf{D} + E_nD_n = -\pi + D_n^2/(2\epsilon) - \epsilon E_s^2/2, \tag{23.4} $$

where E_s is the tangential component of $\mathbf{E}$. At a conductor–dielectric boundary $E_s = 0$ and the continuity of (23.4) again results in the electric force $\frac{1}{2}\sigma\mathbf{E}$. Consider now the boundary between a dielectric and air, for which we set $\chi = 0$ (i.e. $\epsilon = \epsilon_0$). Since D_n and E_s are both continuous, we obtain

$$-\pi + D_n^2/(2\epsilon) - \epsilon E_s^2/2 = -\pi_0 + D_n^2/(2\epsilon_0) - \epsilon_0 E_s^2/2,$$

or

$$\pi_0 - \pi = \tfrac{1}{2}(\epsilon_0^{-1} - \epsilon^{-1})D_n^2 + \tfrac{1}{2}(\epsilon - \epsilon_0)E_s^2, \tag{23.5}$$

where π_0 is the air pressure in the absence of $\boldsymbol{\mathcal{E}}$. Since $\epsilon > \epsilon_0$, the right hand side of (23.5) is positive. Thus the π_0 of the air must exceed the π of the dielectric. If we exhaust the air π_0 vanishes and π must be negative – a tension. This is usually expressed – somewhat loosely – by saying that the right hand side of (23.5) is 'the normal force (per unit area) directed from the dielectric into the vacuum'. Actually, of course, the force is given by (23.4) with the appropriate π and ϵ (0 and ϵ_0 if there is a vacuum outside the dielectric). This pulling of a dielectric fluid towards vacuum is the reason for the rise of a dielectric liquid into a charged capacitor that has been dipped into it.

PROBLEM

Problem 23–1. A conducting sphere of radius a in an infinite medium of uniform permittivity ϵ carries a charge Q. It is divided in two by a diametral plane. Prove that a force of magnitude $Q^2/(32\pi\epsilon a^2)$ is required in order to separate the two halves.

24. Wilson's experiment

Electrostatics does not make full use of the material properties of dielectrics, because the part $\dot{\mathbf{x}} \times \mathbf{B}$ of the electromotive intensity is ignored. If we wish to exhibit the influence of this term, we must examine phenomena in which a dielectric is set in motion (with respect to the aether) and made to cross magnetic field lines. We shall therefore discuss a famous experiment that was designed and conducted by Wilson in 1905 with the object of investigating such an effect. The inner and outer surfaces of a hollow dielectric cylinder (cf. Problem 1–3) are connected through brushes to an electrometer† of capacity C_e. When the cylinder is rotated in a magnetic field $\mathbf{B}_0$ directed along its axis, a voltage is developed between the brushes and the electrometer becomes charged. We wish to calculate this voltage in the steady state that is established after the angular velocity Ω has been maintained constant for some time.‡

The equations of the problem are

$$\operatorname{div} \mathbf{D} = 0, \qquad \mathbf{n} \cdot [\![\mathbf{D}]\!] = \sigma,$$

$$\operatorname{curl} \mathbf{E} = 0, \qquad \mathbf{n} \times [\![\mathbf{E}]\!] = 0,$$

$$\mathbf{D} = \epsilon_0 \mathbf{E} + \mathbf{P}, \qquad \mathbf{P} = \epsilon_0 \chi(\mathbf{E} + \dot{\mathbf{x}} \times \mathbf{B}),$$

$$\operatorname{div} \mathbf{B} = 0, \qquad \mathbf{n} \cdot [\![\mathbf{B}]\!] = 0,$$

$$\operatorname{curl} \mathbf{H} = 0, \qquad \mathbf{n} \times [\![\mathbf{H}]\!] = \mathbf{K},$$

$$\mathbf{H} = \mathbf{B}/\mu_0 + \dot{\mathbf{x}} \times \mathbf{P}. \tag{24.1}$$

Of course, $\dot{\mathbf{x}} = \Omega \times \mathbf{r}$. Using cylindrical coordinates (r, θ, z), we seek a solution with $\mathbf{E} = (E, 0, 0)$ and $\mathbf{B} = (0, 0, B)$. As a result of the equations, we shall then have $\mathbf{P} = (P, 0, 0)$, $\mathbf{D} = (D, 0, 0)$ and $\mathbf{H} = (0, 0, H)$. Since the net charge on the inner and outer surfaces of the dielectric cylinder vanishes, $\mathbf{E}$ will vanish outside the cylinder. Similarly, the convection currents due to the rotating free and bound charges will have no magnetic effect outside: even near the cylinder we shall then have the undistorted magnetic field $\mathbf{B}_0$; but inside $\mathbf{B}$ need not equal $\mathbf{B}_0$.

The equation $\operatorname{div} \mathbf{D} = 0$ gives $rD = \alpha$, a constant. Applying the condition $\mathbf{n} \cdot [\![\mathbf{D}]\!] = \sigma$ at the surfaces, we obtain $\alpha = a\sigma_a = -b\sigma_b$, where σ_a and σ_b are the densities of the surface charges deposited there by the electrometer. Thus, for $a < r < b$,

$$D = \epsilon_0 E + P = \epsilon_0 E + \epsilon_0 \chi(E + \Omega r B) = \frac{a}{r}\sigma_a, \tag{24.2}$$

and everywhere else $D = 0$.

---

† An electrometer is a capacitor with a measuring device for the force between its electrodes; it is used for measuring a potential difference.
‡ In §32 we shall see that this time is practically zero.

Turning now to the magnetic field, **curl H** = 0 states that

$$H = B/\mu_0 - \Omega r P = B/\mu_0 - \Omega r \epsilon_0 \chi (E + \Omega r B)$$

is constant. Applying the condition **n** × [[**H**]] = **K** to the surfaces, we obtain $B_0/\mu_0 - H = \Omega a \sigma_a = -\Omega b \sigma_b$. Hence

$$(1 - \chi \Omega^2 r^2/c^2) B/\mu_0 = B_0/\mu_0 - \Omega a \sigma_a + \Omega r \epsilon_0 \chi E. \tag{24.3}$$

Eliminating B between (24.2) and (24.3), and neglecting the small (relativistic) term $\chi \Omega^2 r^2/c^2$ everywhere, we find

$$\epsilon E = \frac{a}{r} \sigma_a - (\epsilon - \epsilon_0) \Omega B_0 r,$$

where, as usual, $\epsilon = (1 + \chi)\epsilon_0$. Since $E = -dV/dr$ (that is where **curl E** = 0 is used), this integrates to

$$\epsilon v = -a \sigma_a \ln \frac{b}{a} + \tfrac{1}{2}(\epsilon - \epsilon_0) \Omega B_0 (b^2 - a^2),$$

where $v = V_b - V_a$ is the voltage. Using the fact that the charge $2\pi a L \sigma_a$ deposited by the electrometer on the inner surface is $C_e v$, we finally obtain

$$v = \frac{\epsilon - \epsilon_0}{\epsilon} \frac{\Omega}{2\pi} \frac{\pi(b^2 - a^2)B_0}{1 + C_e/C}, \tag{24.4}$$

where C (cf. (22.11)) denotes the capacity of the cylinder. We note that $\Omega/(2\pi)$ is the rate of rotation in revolutions per unit time and that $\pi(b^2 - a^2)$ is the cross section of the cylinder; the product $\pi(b^2 - a^2)B_0$ is therefore the flux through the cylinder of the external magnetic field.

Wilson's experiment confirmed the linear dependence of the voltage on Ω or B_0. But its historical importance was in confirming the way in which the v of (24.4) depends on the dielectric susceptibility χ. At the turn of the century there was a great controversy regarding the motion of bodies through the aether. According to the principles we have adopted, $\mathbf{P} = (\epsilon - \epsilon_0)\boldsymbol{\mathcal{E}}$ and $\mathbf{D} = \epsilon_0 \mathbf{E} + \mathbf{P} = \epsilon_0 \mathbf{E} + (\epsilon - \epsilon_0)\boldsymbol{\mathcal{E}}$. This was the view of Lorenz. Hertz, on the other hand, advocated a different view, based on the belief that bodies with electromagnetic properties carried the aether along with them. According to Hertz, the relation between the electric displacement and the electromotive intensity in a moving linear dielectric was $\mathbf{D} = \epsilon\boldsymbol{\mathcal{E}}$. Hertz's relation would have led, in Wilson's experiment, to a voltage that was *independent* of the dielectric permittivity. Instead, the experiment confirmed the prediction of Lorenz, that v should vary as $(\epsilon - \epsilon_0)/\epsilon = \chi/(1 + \chi)$.

PROBLEM

Problem 24–1. Find the potential outside a dielectric sphere of radius a and uniform permittivity ϵ which is placed in a uniform magnetic field **B** and made to rotate at constant angular velocity Ω around an axis parallel to **B**.

25. Pyroelectricity and piezoelectricity

We have seen that the electrostatics of linear, isotropic, dielectrics with constant permittivity is governed by Laplace's equation. The ensuing potential problems are similar to those which we have encountered with conductors in vacuum and can be tackled by the same methods. As in the case of conductors, the electric part of the stress is quadratic in $\mathcal{E}$. If the susceptibility χ is variable, or – in the case of non-isotropic linear dielectrics – if it is a matrix, the equations and the boundary conditions are still linear and therefore relatively simple to handle, and the electric part of the stress still quadratic in $\mathcal{E}$.

Not all dielectrics are so obliging, however. In some of them the polarization **P** is not only non-linear in the electromotive intensity $\mathcal{E}$: it even depends on the history of the material. Such are the *ferroelectrics*, which we shall discuss in the next section. The history dependence, called *hysteresis*, means that they can *not* be described by the material classes of §15–16. Methods have been developed for dealing with the thermodynamics of materials with memory, but they lie outside the scope of this book. (Of course they are still based on the principles we have laid down, in particular on the entropy inequality (14.11), but they deal with classes of materials more general than the ones defined by (15.1) or (16.1).)

Whether they are ferroelectric or not, dielectrics exist that may be polarized even in the absence of any electromotive intensity. Such materials are the electric analogues of permanent magnets. They are called *electrets*, and when they are in a state, or phase, with $\mathcal{E} = 0$ and $\mathbf{P} \neq 0$, the state (or phase) is said to be *pyroelectric*.

Consider, for example, a pyroelectric sphere of radius a which is uniformly polarized. We wish to determine the electric field outside the sphere (where we assume a vacuum, or air, with $\epsilon = \epsilon_0$.) Obviously, we must solve a potential problem with the jump conditions $\mathbf{n} \times [\![\mathbf{E}]\!] = 0$ and $\mathbf{n} \cdot [\![\mathbf{D}]\!] = 0$ on $r = a$. This can be done by expanding the potential as a series in $r^n P_n(\cos\theta)$ inside and $r^{-(n+1)} P_n(\cos\theta)$ outside (with θ measured from the direction of $\mathbf{P}$). We prefer a short cut. Since $\mathbf{P}$ is uniform, the bound-charge density $q_R = -\operatorname{div}\mathbf{P}$ vanishes inside the sphere, but there is a surface density of bound charge, $\sigma_R = \mathbf{P} \cdot \mathbf{n}$. Far from the sphere, the electric field must therefore be a dipole field of the form (20.9). Let us assume that the external field $\mathbf{E}_e$ has this form even near the sphere and write

$$\epsilon_0 \mathbf{E}_e = \alpha \frac{a^3}{r^3}[3(\mathbf{P} \cdot \mathbf{n})\mathbf{n} - \mathbf{P}], \tag{25.1}$$

where α is a constant and $\mathbf{n}$ is a unit vector in the direction of the radius-vector from the centre of the sphere. Inside, we assume a uniform field, say

$$\epsilon_0 \mathbf{E}_i = \beta\mathbf{P}, \tag{25.2}$$

where β is another constant. The condition $\mathbf{n} \times [\![\mathbf{E}]\!] = 0$ gives $(\alpha + \beta)\mathbf{n} \times \mathbf{P} = 0$, or $\alpha + \beta = 0$. The condition $\mathbf{n} \cdot [\![\mathbf{D}]\!] = 0$ gives $(2\alpha - 1 - \beta)(\mathbf{P} \cdot \mathbf{n}) = 0$ or $2\alpha - 1 - \beta = 0$. Thus $\alpha = 1/3$ (and $\beta = -1/3$). We have found a solution. It is easy to check that there is no other. Hence it is *the* solution. The external field is therefore

$$\mathbf{E}_e = \frac{1}{4\pi\epsilon_0} \frac{3(\boldsymbol{\mathcal{P}} \cdot \mathbf{n})\mathbf{n} - \boldsymbol{\mathcal{P}}}{r^3}, \tag{25.3}$$

where $\boldsymbol{\mathcal{P}} = \int \mathbf{P}\, dV$ is the dipole moment of the pyroelectric sphere. Inside, the electric field is $\mathbf{E}_i = -\mathbf{P}/(3\epsilon_0)$. It is uniform and independent of the size of the sphere.

PROBLEM

Problem 25–1. Determine the surface force on a pyroelectric.

We have already noted that the material classes we have introduced for fluids and elastic materials in §15–16 cannot describe ferroelectric behaviour, since they assume that all properties depend on the present values of the arguments. They can, however, be used to illustrate pyroelectricity, as well as another phenomenon – *piezoelectricity*. Piezoelectricity is a property of many crystals – quartz being the notorious example – in which polarization arises as a result of deformation, even in the absence of $\mathcal{E}$, and stresses appear that are linear, rather than quadratic, in $\mathcal{E}$.

In a dielectric which is an elastic material, the free energy must be of the form

$$\varphi = \Phi(F^T F, \vartheta, F^T \mathcal{E}). \tag{25.4}$$

This is the reduced form (16.11) that satisfies both the entropy inequality and Euler's second law. We have left out the dependence on $\mathbf{B}$, since $\mathcal{M} = -\rho\varphi_\mathbf{B} = 0$. The function Φ depends on the six independent components of the symmetric matrix $F^T F$, on the temperature ϑ, and on the three components of $F^T \mathcal{E}$. The polarization and stress involve the derivatives $\varphi_\mathcal{E}(F, \vartheta, \mathcal{E})$ and $\varphi_F(F, \vartheta, \mathcal{E})$, which are connected with the derivatives of Φ through the chain rule of differentiation. It is easy to show that

$$\varphi_\mathcal{E} = F\Phi_{F^T \mathcal{E}},$$

$$\varphi_F = 2F\Phi_{F^T F} + \mathcal{E} \otimes \Phi_{F^T \mathcal{E}}. \tag{25.5}$$

In terms of the derivatives of Φ, the polarization (16.4) and the stress (16.6) in an elastic dielectric become

$$\mathbf{P} = -\rho F\Phi_{F^T \mathcal{E}}, \tag{25.6}$$

$$T = 2\rho F\Phi_{F^T F} F^T - \tfrac{1}{2}\epsilon_0 E^2 I + \epsilon_0 \mathbf{E} \otimes \mathbf{E}. \tag{25.7}$$

Consider now a dielectric described by a Φ which is of second degree in the electromotive intensity:

$$\Phi = a_i(F^T \mathcal{E})_i + b_{ij,k}(F^T F)_{ij}(F^T \mathcal{E})_k - \tfrac{1}{2}\epsilon_0 c_{ij}(F^T \mathcal{E})_i(F^T \mathcal{E})_j, \tag{25.8}$$

where the coefficients a, b and c are functions of the temperature; there is no restriction in assuming $b_{ij,k}$ and c_{ij} to be symmetric in i and j (in the case of the former these indices are separated from the third by a comma). If the electric field is weak and the deformation small (F being close to the identity I), the dependence of Φ on $\mathcal{E}$ is well approximated by (25.8). Thus our Φ is more than a prelude to a mathematical game.

Since the polarization and stress of (25.6)–(25.7) are linear in the derivatives of Φ, each of the three terms of (25.8) makes its individual contribution. The first and third do not contribute to T because they do not involve $F^T F$. They give rise to the following polarization:

$$\mathbf{P} = -\rho F \mathbf{a} + \epsilon_0 \rho F c F^T \mathcal{E}. \tag{25.9}$$

The first term on the right hand side of this equation is independent of $\mathcal{E}$ and therefore pyroelectric. The second is linear in $\mathcal{E}$, with a symmetric susceptibility matrix $\chi = \rho F c F^T$.

The second term of (25.8) involves both $F^T \mathcal{E}$ and $F^T F$. It gives rise to a polarization which depends on the deformation F, but not on $\mathcal{E}$, and also to a stress which is linear in $\mathcal{E}$. Both effects are controlled by the same coefficients $b_{ij,k}$ (eighteen at most). This is piezoelectricity. In a sufficiently weak $\mathcal{E}$ the linear, piezoelectric term in the stress will dominate the other, quadratic terms. Conversely, the electromotive intensity in a piezoelectric material varies as the stress, rather than as its square root. In piezoelectric cigarette lighters this effect is used in order to produce a spark by squeezing.

Other interesting phenomena in piezoelectrics are connected with oscillations and the propagation of elastic waves. They do not belong to electrostatics. Rather, they are governed by the following system of equations:

$$\operatorname{div} \mathbf{D} = 0, \qquad \operatorname{curl} \mathbf{H} = \mathbf{D}_t,$$

$$\mathbf{D} = \epsilon_0 \mathbf{E} + \mathbf{P},$$

$$\operatorname{div} \mathbf{B} = 0, \qquad \operatorname{curl} \mathbf{E} = -\mathbf{B}_t,$$

$$\mathbf{H} = \mathbf{B}/\mu_0 + \dot{\mathbf{x}} \times \mathbf{P},$$

$$\rho \ddot{\mathbf{x}} = \operatorname{div} T + \rho \mathbf{b}. \tag{25.10}$$

To this system we must add the corresponding jump conditions and substitute the expressions for $\mathbf{P}$ and T that follow from the middle, piezoelectric term of (25.8); for example, $P_i = -\rho F_{ij} b_{nm,j} (F^T F)_{nm}$.

Obviously, the system is far from simple. Progress can be made by linearization, i.e. by casting away all terms that are quadratic in $\mathbf{E}$, $\mathbf{B}$, $\dot{\mathbf{x}}$ or $F - I$; often the magnetic field is left out entirely.

The effects of pyroelectricity, linear polarization and piezoelectricity are thus seen to arise quite naturally in elastic dielectrics. In fact, we should expect them to occur simultaneously. In a very weak field the first two terms of (25.8) are the most important, and the material will exhibit pyroelectric or piezoelectric behaviour. In stronger fields the third term of (25.8) will dominate, and we expect the same material to be a linear dielectric. It is therefore not surprising that tables of material permittivities list quartz, a common piezoelectric, as having $\epsilon = 3.85\epsilon_0$. Finally, since the coefficients a, b and c are functions of the temperature, we also expect the magnitudes of these effects to be temperature-dependent.

26. Ferroelectrics

Ferroelectrics are materials in which the permanent polarization can be changed – and even reversed – by an electric field. They are the electric analogues of ferromagnets (but, unlike the latter, they have nothing to do with iron). Like ferromagnets, they exhibit non-linear behaviour and hysteresis. They may be strongly polarized even in the absence of any electromotive intensity, but only so long as the temperature lies below a characteristic temperature ϑ_c, called the *Curie temperature*. Some of these properties can be shown to follow from a suitable choice† of the free energy φ, which determines the polarization through $\mathbf{P} = -\rho\varphi_{\mathcal{E}}$.

In order to keep things as simple as possible, we consider an electric material from the simple class of fluids (§15) for which φ depends on ρ, ϑ and $\mathcal{E}$. Obviously, we cannot hope to exhibit the phenomenon of hysteresis in this way, since the φ's we shall be dealing with depend on *present* values of their arguments.

We now regard $\rho\varphi_{\mathcal{E}}(\rho, \vartheta, \mathcal{E}) = -\mathbf{P}$ as an equation that (implicitly) determines $\mathcal{E}$ as a function of ρ, ϑ and $\mathbf{P}$, and then carry out a Legendre transformation (like the transformation from a Lagrangian to a

† Due to V. L. Ginzburg.

Hamiltonian) to a new thermodynamic potential:

$$\rho\tilde{\varphi}(\rho,\vartheta,\mathbf{P}) = \rho\varphi + \mathbf{P}\cdot\boldsymbol{\mathcal{E}}. \qquad (26.1)$$

Since $\boldsymbol{\mathcal{E}}$ on the right hand side is now regarded as the solution of $\rho\varphi_{\mathcal{E}} = -\mathbf{P}$, the new thermodynamic potential $\tilde{\varphi}$ depends on ρ, ϑ and $\mathbf{P}$. Its $\mathbf{P}$-derivative is easily obtained by the chain rule:

$$\rho\tilde{\varphi}_{\mathbf{P}}(\rho,\vartheta,\mathbf{P}) = \boldsymbol{\mathcal{E}}. \qquad (26.2)$$

Instead of specifying an electric material by a $\varphi(\rho,\vartheta,\boldsymbol{\mathcal{E}})$, we can specify it by a $\tilde{\varphi}(\rho,\vartheta,\mathbf{P})$; this procedure is, of course, analogous to specifying a dynamical system by a Hamiltonian rather than by a Lagrangian.

Consider now the thermodynamic potential

$$\rho\tilde{\varphi}(\rho,\vartheta,\mathbf{P}) = \rho\tilde{\varphi}_0(\rho,\vartheta) + [a(\vartheta-\vartheta_c)P^2 + bP^4]/\epsilon_0, \qquad (26.3)$$

where a and b are both positive and independent of ϑ or $\mathbf{P}$, and ϑ_c is a positive constant. We note that $\tilde{\varphi}$ is rotationally invariant, since it depends only on the magnitude of $\mathbf{P}$. The same will be true of φ, which according to (26.1) differs from $\tilde{\varphi}$ by a scalar product of vectors. This ensures that φ will obey the restriction imposed by Euler's second law of mechanics (cf. the statement following (15.17)). Substitution of (26.3) in (26.2) gives

$$[2a(\vartheta-\vartheta_c) + 4bP^2]\mathbf{P} = \epsilon_0\boldsymbol{\mathcal{E}}. \qquad (26.4)$$

The polarization $\mathbf{P}$ is parallel or anti-parallel to $\boldsymbol{\mathcal{E}}$, depending on the sign of the expression in the square brackets of (26.4). This is, of course, a consequence of the isotropy of the particular $\tilde{\varphi}$ we have chosen.

If $\vartheta > \vartheta_c$, $\mathbf{P}$ is always parallel to $\boldsymbol{\mathcal{E}}$. When $\boldsymbol{\mathcal{E}}$ is small, the relationship is linear, with a susceptibility given by

$$\chi = \frac{(2a)^{-1}}{\vartheta-\vartheta_c}. \qquad (26.5)$$

This behaviour (called *paraelectric*, in analogy with the behaviour that is called *paramagnetic* in the magnetic case), with a susceptibility χ inversely proportional to $\vartheta - \vartheta_c$, is the (ferroelectric) *Curie-Weiss law*. It is rather well substantiated by experiments. The critical temperature

ϑ_c is the (ferroelectric) *Curie temperature* or the *Curie point*. When $\mathcal{E}$ is very large, (26.4) gives

$$\mathbf{P} = (4b\epsilon_0^2 \mathcal{E}^2)^{-1/3} \epsilon_0 \mathcal{E}. \tag{26.6}$$

In actual ferroelectrics $\mathbf{P}$ attains a constant *saturation* value for very large $\mathcal{E}$. Equation (26.6) does not lead to saturation, although it does predict that $\mathbf{P}$ will increase more slowly than $\mathcal{E}$. It is therefore not valid for very large $\mathbf{P}$. This failure is not wholly unexpected, since the $\tilde{\varphi}$ of (26.3) looks like the beginning of an expansion in powers of P^2. We should not expect it to hold when $\mathbf{P}$ is large.

Below the Curie temperature the $\mathbf{P}(\mathcal{E})$ relationship is more complicated. For $\mathcal{E} = 0$ we have, besides $\mathbf{P} = 0$, the *spontaneous* ferroelectric solution

$$P^2 = \frac{a(\vartheta_c - \vartheta)}{2b}. \tag{26.7}$$

Since it corresponds to the vanishing of the square bracket in (26.4), its direction is undetermined. (Again, this is a consequence of the isotropy of (26.3).) For $\mathcal{E} \neq 0$, every solution has a determinate direction, and we may assume that $\mathbf{P}$ and $\mathcal{E}$ both lie on the z axis:

$$[2a(\vartheta - \vartheta_c) + 4bP_z^2]P_z = \epsilon_0 \mathcal{E}_z. \tag{26.8}$$

The resulting polarization curve is shown in Figure 7.

Clearly,

$$[2a(\vartheta - \vartheta_c) + 12bP_z^2]\frac{\partial P_z}{\partial \mathcal{E}_z} = \epsilon_0. \tag{26.9}$$

The extrema c and d are the points at which $\partial \mathcal{E}_z / \partial P_z = 0$:

$$P_z^2 = \frac{a(\vartheta_c - \vartheta)}{6b}. \tag{26.10}$$

For each value of $\mathcal{E}_z$ between the abscissae corresponding to c and d the cubic equation (26.8) has three real roots. One of these must, however, lie on the dashed part cd of the curve, which turns out to correspond to unstable states. In order to see this, we substitute (26.3) into (26.1) and obtain

$$\rho\varphi = \rho\tilde{\varphi}(\rho, \vartheta) + [a(\vartheta - \vartheta_c)P_z^2 + bP_z^4]/\epsilon_0 - P_z\mathcal{E}_z. \tag{26.11}$$

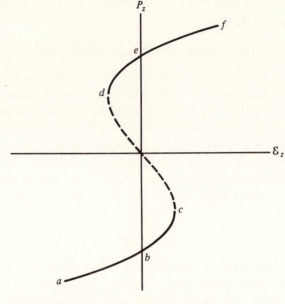

FIGURE 7

Equation (26.8) is the same as $\partial\varphi/\partial P_z = 0$, the derivative being taken at fixed ρ, ϑ and $\mathcal{E}_z$. The polarization curve is thus the locus of extrema of φ. Equation (26:9) is the same as

$$\left(\rho\frac{\partial^2\varphi}{\partial P_z^2}\right)\frac{\partial P_z}{\partial \mathcal{E}_z} = \epsilon_0; \tag{26.12}$$

the two factors on the left hand side must therefore have the same sign. This shows that, wherever $\partial P_z/\partial\mathcal{E}_z > 0$, φ has a minimum, and wherever $\partial P_z/\partial\mathcal{E}_z < 0$, φ has a maximum. The dashed part cd of the curve therefore corresponds to states which are unstable (cf. §14).

We have shown that, for $\mathcal{E}_z$ between the abscissae corresponding to c and d, there are only two stable solutions, corresponding to minima of φ. They give rise to oppositely directed polarizations. Above the Curie temperature these ferroelectric solutions disappear, and the material becomes paraelectric.

The thermodynamic potentials we have considered above were based on the class of simple fluids. Since real ferroelectric materials are crystalline or polycrystalline solids, we might (at least) try to construct free energies based on the reduced form (25.4) for elastic materials. The resulting expressions will be quite involved, and we shall not attempt to derive them. Even so, we may still expect them to lead to non-zero ferroelectric solutions (even when $\mathcal{E} = 0$) below a definite temperature. But in these solutions $\mathbf{P}$ will generally have components that are *transverse* with respect to $\mathcal{E}$, and the $\mathbf{P}(\mathcal{E})$ relation will be described by a polarization *surface* (or hyper-surface) rather than by a polarization *curve*. In the simplest case, the material has a single ferroelectric *axis*, and the ferroelectric solutions can only have one of the two directions defined by this axis. Usually there are several such axes, but even if they are all equivalent – as in a crystal of cubic symmetry – the establishment of polarization along one of them will result in distortion due to electrostriction (i.e. electric stresses). This will cause some of the axes to become 'easy' and others 'hard'; these terms refer, of course, to the components of $\mathbf{P}$ that arise in response to components of $\mathcal{E}$ along these axes.

In addition to the $\mathbf{P}(\mathcal{E})$ relation, $\mathbf{P}$ and $\mathcal{E}$ must also satisfy the electromagnetic, mechanical and thermodynamic equations and jump conditions. In an electrostatic situation, these are

$$\operatorname{div}(\epsilon_0 \mathbf{E} + \mathbf{P}) = 0, \quad \mathbf{n} \cdot [\![\epsilon_0 \mathbf{E} + \mathbf{P}]\!] = 0,$$

$$\operatorname{curl} \mathbf{E} = 0, \quad \mathbf{n} \times [\![\mathbf{E}]\!] = 0, \tag{26.13}$$

as well as the equation and jump condition of mechanical equilibrium. Taken together, these conditions turn out to be quite stringent. Imagine, for example, a sphere of uniaxial ferroelectric material which has been cooled down through its Curie temperature in the absence of an applied electric field. We choose the ferromagnetic axis as the z axis and continue to apply our $\tilde{\varphi}$ of (26.3) with the understanding that $\mathbf{P} = (0, 0, P_z)$. When $\vartheta < \vartheta_c$, the material will become spontaneously polarized, with P_z equal to one of the roots of (26.7). If P_z has the same sign everywhere, we have the uniform pyroelectric sphere for which the equations (26.13) were solved in §25. According to (25.2), there will be a uniform field

$\mathbf{E} = -\mathbf{P}/(3\epsilon_0)$ throughout the sphere (it is often referred to as the *depolarizing field*) which is independent of the sphere's size. Now the polarization curves of all known ferroelectrics are such that this opposite field is orders of magnitude greater than the field required for a reversal of P_z. Of course, a mere flipping over of the polarization will leave the ferroelectric sphere in the same predicament. We must therefore conclude that P_z cannot have the same sign throughout the sphere. This is supported by observations: the ferromagnetic substance divides itself into regions or *domains* of uniform polarization in alternating directions. The shapes and sizes of these domains depend on the geometry of the sample, on the temperature and on the applied stresses and electric fields. The domain boundaries are observed to move – thus changing the domain structure – in response to changes in these parameters.

The manner in which domain formation reduces the depolarizing fields can be illustrated by the study of simple domain structures. Consider a layer of thickness ℓ made of uniaxial ferroelectric material, with the axis perpendicular to the layer. Figure 8 shows two possible domain structures.

We shall calculate the electric field for the striped structure. Since $\mathbf{P}$ is uniform (except for jumps across the domain walls), equations (26.13) reduce to a potential problem for the electric field. Clearly, P_z is a periodic function with period $2d$. We choose the upper face as the plane $z = 0$, and the x-coordinate such that $P_z = -P$ for $-d < x < 0$ and $P_z = +P$ for $0 < x < d$. The Fourier expansion of P_z is then

$$P_z(x) = \sum_{n=0}^{\infty} b_n \sin \frac{(2n+1)\pi x}{d}, \qquad b_n = \frac{4P}{(2n+1)\pi}. \qquad (26.14)$$

For the potential, we seek a solution of $V_{xx} + V_{zz} = 0$ which has the form

$$V(x,z) = \sum f_n(z) \sin \frac{(2n+1)\pi x}{d} \qquad (26.15)$$

and remains finite as we move away from $z = 0$. This gives

$$f_n(z) = c_n e^{\mp(2n+1)\pi z/d}, \qquad (26.16)$$

the signs in the exponent referring to $z > 0$ and $z < 0$. The c_n are

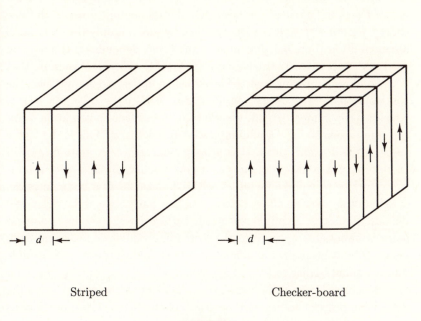

Striped   Checker-board

FIGURE 8

related to the b_n of (26.14) by the jump condition $(26.13)_2$:

$$c_n = \frac{d}{2\pi\epsilon_0(2n+1)}b_n. \tag{26.17}$$

The electric field $\mathbf{E} = -\operatorname{\mathbf{grad}} V$ therefore decays exponentially away from the layer face. Inside the material it is directed so that $\mathbf{P} \cdot \mathbf{E} < 0$.

Having understood the necessity for domain formation, we are now led to ask: why does the subdivision into domains ever cease, rather than go on indefinitely, perhaps down to atomic sizes? In order to account for this, theories of ferroelectricity postulate a surface tension along the domain walls. This means the addition, to the free energy, of a surface term $\int \alpha \, dS$, where the *surface tension coefficient* α is positive; presumably it depends on the temperature, the deformation, the polarization, etc.

We may illustrate the effect of this term with the foregoing example of striped domain structure in a layer. Since the area of each domain wall is ℓ times the depth of the layer, the term $\int \alpha \, dS$ makes a contribution of $\alpha\ell/d$ per unit area of the layer. Of the remaining terms in the free energy $\int \varphi \, dm = \int \rho\varphi \, dV$, with $\rho\varphi$ given by (26.11), only the last makes a contribution that, per unit area of the layer, depends on the domain thickness d. It is easily calculated from (26.14)–(26.17). Assuming that $d \ll \ell$ and remembering that the layer has an upper and a lower face, the contribution of $\int -P_z E_z \, dV$, per unit area, is $4P^2 d/(\pi^3\epsilon_0) \sum (2n+1)^{-3}$, which is proportional to d. If we add to it the surface contribution $\alpha\ell/d$, the sum is seen to have a minimum with respect to d. This is the stable value of the domain width. It is proportional to $\sqrt{\ell}$. A finite block will therefore contain fewer domains as it gets smaller. In fact, since the ratio of surface to volume increases as the dimensions of a body decrease, it follows that very small ferroelectric bodies must cease to be polarized: domain formation requires too much surface energy; a single-domain state is impossible because of the forbiddingly high depolarizing field; hence $\mathbf{P} = 0$ becomes a stable state. It should, however, be possible for very small bodies to form single domains if the surrounding medium is conducting, because the depolarizing field would be cancelled by the deposition of free charges on the surface of the body. These predictions are all borne out by observations.

CHAPTER VIII

Magnetism

27. Magnetic effects of currents

In the absence of polarization and magnetization, $\mathbf{H} = \mathbf{B}/\mu_0$, and the equations governing a steady system of currents are

$$\operatorname{div} \mathbf{B} = 0,$$

$$\operatorname{curl} \mathbf{B} = \mu_0 \mathbf{j}. \tag{27.1}$$

If we regard $\mathbf{j}$ as given, these equations determine $\mathbf{B}$. The first is satisfied by introducing a vector potential

$$\mathbf{B} = \operatorname{curl} \mathbf{A}. \tag{27.2}$$

The potential $\mathbf{A}$ is subject to a gauge transformation, and this can be chosen in such a way as to impose the condition

$$\operatorname{div} \mathbf{A} = 0. \tag{27.3}$$

Such a vector potential is said to satisfy the *Coulomb gauge*.

If we now combine (27.2) and (27.3) with the remaining equation $(27.1)_2$ we obtain the equation

$$\Delta \mathbf{A} = -\mu_0 \mathbf{j}, \tag{27.4}$$

where $\Delta \mathbf{A} = \operatorname{grad} \operatorname{div} \mathbf{A} - \operatorname{curl}^2 \mathbf{A}$. The Cartesian components of $\Delta \mathbf{A}$ are $(\Delta A_x, \Delta A_y, \Delta A_z)$. The Cartesian components of (27.4) are therefore three Poisson's equations. Hence the regular solution of (27.4) is (cf. (19.17))

$$A = \frac{\mu_0}{4\pi} \int \frac{\mathbf{j} d^3 x}{r}, \tag{27.5}$$

where r is the distance from the volume element d^3x to the point at which $\mathbf{A}$ is evaluated. In order to get $\mathbf{B}$, we must take the curl; note that the curl only operates on r. The result is

$$\mathbf{B} = \frac{\mu_0}{4\pi} \int \mathbf{grad} \frac{1}{r} \times \mathbf{j} d^3 x = -\frac{\mu_0}{4\pi} \int \frac{\mathbf{r} \times \mathbf{j} d^3 x}{r^3}. \tag{27.6}$$

For linear currents we replace the vector $\mathbf{j} d^3 x$ by the vector $i\mathbf{ds}$, where $\mathbf{ds}$ is an element of the line, directed parallel to the current:

$$\mathbf{A} = \frac{\mu_0 i}{4\pi} \oint \frac{\mathbf{ds}}{r}, \qquad \mathbf{B} = \frac{\mu_0 i}{4\pi} \oint \frac{\mathbf{ds} \times \mathbf{r}}{r^3}. \tag{27.7}$$

Equation $(27.7)_2$ is the formula of Biot and Savart. Since $\mathbf{H} = \mathbf{B}/\mu_0$, it expresses the current potential of linear currents as

$$\mathbf{H} = \frac{i}{4\pi} \oint \frac{\mathbf{ds} \times \mathbf{r}}{r^3}. \tag{27.8}$$

The integrals in (27.7) extend over all current lines. If there are several current loops, each integral is a sum.

PROBLEM

Problem 27–1. A current i flows in a circular wire of radius a, as in Figure 9. Show that at a point P on the axis $\mathbf{H}$ is directed along the axis and has the magnitude

$$H_P = \frac{1}{2} i \frac{a^2}{(a^2 + z^2)^{3/2}}. \tag{27.9}$$

With linear currents the system (27.1) is homogeneous, except for singularities of **curl** $\mathbf{H}$ on the lines along which currents are flowing. It is then possible to introduce a scalar magnetic potential Ω, such that

$$\mathbf{H} = -\mathbf{grad}\,\Omega, \qquad \Delta\Omega = 0. \tag{27.10}$$

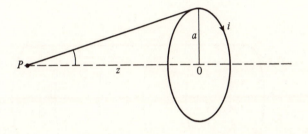

FIGURE 9

PROBLEM

Problem 27–2. For a single current loop, use (27.8) and Stokes's theorem to prove that a scalar potential at P is provided by

$$\Omega_P = \frac{i}{4\pi}\omega_P, \tag{27.11}$$

where ω_P is the solid angle subtended by the loop at P.†

If we apply (27.11) to the current loop of Figure 9, we obtain

$$\Omega_P = \tfrac{1}{2}i(1 - \cos\theta). \tag{27.12}$$

From this and $H = -\partial\Omega/\partial z$ we arrive again at (27.9).

---

† The potential Ω changes sign on passing from one side of the loop to the other. This is because the orientation of the surface in Stokes's theorem is determined by the sense in which the loop is traversed. This sense is fixed by the direction of the current, but $d\omega = (\mathbf{r} \cdot \mathbf{n})\,dS/r^3$ changes sign along with $\mathbf{r} \cdot \mathbf{n}$ on passing from one side to the other.

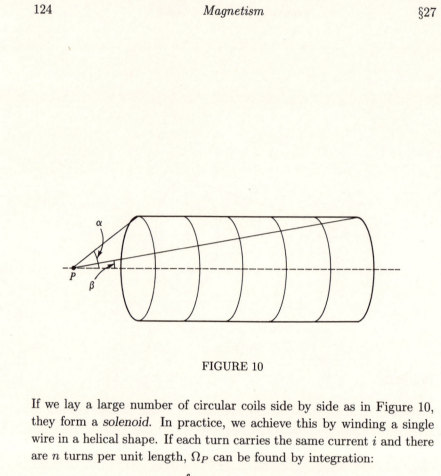

FIGURE 10

If we lay a large number of circular coils side by side as in Figure 10, they form a *solenoid*. In practice, we achieve this by winding a single wire in a helical shape. If each turn carries the same current i and there are n turns per unit length, Ω_P can be found by integration:

$$\Omega_P = \int \tfrac{1}{2}i(1 - \cos\theta)n\,dz$$

$$= -\tfrac{1}{2}ni\int_\alpha^\beta (1 - \cos\theta)a\sin^{-2}\theta\,d\theta = \tfrac{1}{2}nia(\tan\tfrac{1}{2}\alpha - \tan\tfrac{1}{2}\beta). \quad (27.13)$$

From this we obtain

$$H = -\frac{\partial\Omega}{\partial z} = \tfrac{1}{2}ni(\cos\beta - \cos\alpha). \quad (27.14)$$

If the solenoid is infinite, we set $\alpha = \pi$ and $\beta = 0$. Then

$$H = ni. \quad (27.15)$$

We have seen (cf. (15.25)) that in the present case of vanishing polarization and magnetization it is possible to define a magnetic energy as

$$U = \int \frac{B^2}{2\mu_0} \, d^3x = \tfrac{1}{2} \int \mathbf{H} \cdot \mathbf{B} \, d^3x. \tag{27.16}$$

Now div $\mathbf{A} \times \mathbf{H} = \mathbf{H} \cdot \mathbf{curl}\ \mathbf{A} - \mathbf{A} \cdot \mathbf{curl}\ \mathbf{H} = \mathbf{H} \cdot \mathbf{B} - \mathbf{A} \cdot \mathbf{j}$. If we integrate this and note that the surface integral at infinity vanishes in accordance with the behaviour of $\mathbf{A}$ and $\mathbf{H}$ (cf. $(27.7)_1$ and (27.8)), we get

$$U = \tfrac{1}{2} \int \mathbf{A} \cdot \mathbf{j} \, d^3x = \tfrac{1}{2} i \oint \mathbf{A} \cdot \mathbf{ds}. \tag{27.17}$$

We denote by

$$\mathbf{A}_a = \frac{\mu_0}{4\pi} i_a \oint \frac{\mathbf{ds}_a}{r} \tag{27.18}$$

the vector potential due to the ath current. Of course $\mathbf{A} = \sum \mathbf{A}_a$. Then

$$U = \tfrac{1}{2} \sum i_a \oint \mathbf{A} \cdot \mathbf{ds}_a = \frac{\mu_0}{8\pi} \sum i_a i_b \oint \oint \frac{\mathbf{ds}_a \cdot \mathbf{ds}_b}{r}$$

$$= \tfrac{1}{2} \sum L_{ab} i_a i_b, \tag{27.19}$$

where the last two sums are, of course, double sums. The coefficients

$$L_{ab} = \frac{\mu_0}{4\pi} \oint \oint \frac{\mathbf{ds}_a \cdot \mathbf{ds}_b}{r} \tag{27.20}$$

are the *coefficients of inductance*. Since $U > 0$, the matrix L is positive definite. In particular, the coefficients of *self-inductance* L_{aa} are all positive, and the coefficients of *mutual inductance* L_{ab} satisfy $L_{ab}^2 \le L_{aa}L_{bb}$. Another way of writing U is

$$U = \sum \tfrac{1}{2} i_a \oint \mathbf{A} \cdot \mathbf{ds}_a$$

$$= \sum \tfrac{1}{2} i_a \int (\mathbf{curl}\ \mathbf{A})_n \, dS_a = \tfrac{1}{2} \sum i_a \Phi_a, \tag{27.21}$$

where

$$\Phi_a = \int B_n \, dS_a \tag{27.22}$$

is the magnetic flux through the ath circuit. Clearly

$$\Phi_a = \sum L_{ab} i_b. \tag{27.23}$$

Mutual inductance coefficients are usually calculated from this formula, rather than from the Helmholtz formula (27.20).

The units of L_{ab} are those of magnetic flux, divided by current. The SI unit weber/amp is called the *henry*.

PROBLEMS

Problem 27–3. A current i is flowing along a circular wire of radius a. Find the scalar magnetic potential Ω (at points on or off the axis) by using the result of Problem 27–1 and the device of Problem 21–6.

Problem 27–4. Two coils, one of m turns and one of n turns, are wound on a circular ring of arbitrary cross section. Show that the coefficients of inductance are

$$L_{11} = \frac{m^2}{2\pi} \int \frac{dS}{s}, \qquad L_{22} = \frac{n^2}{2\pi} \int \frac{dS}{s}, \qquad L_{12} = \frac{mn}{2\pi} \int \frac{dS}{s},$$

where s is the distance of the element dS of the cross section from the axis of the ring. Note that $L_{11} L_{22} = L_{12}^2$.

28. Magnetic materials

Magnetic materials, or magnets, are analogous to dielectrics and present similar phenomena. The *ferromagnets* show hysteresis, sometimes with dramatic temperature dependence. We shall deal with some aspects of ferromagnetism in the next section. If, on the other hand, the state of a magnet depends only on present values, we may apply the theory of §15–16. There is nothing, in principle, to prevent a material from being both dielectric and magnetic. Common magnets, however, show no polarization. Then $\mathbf{P} = -\rho \varphi_{\mathcal{E}} = 0$, so that the free energy is independent of the electromotive intensity $\mathcal{E}$. The same is then true of the magnetization

$$\mathcal{M} = -\rho \varphi_{\mathbf{B}}. \tag{28.1}$$

Linear magnets are materials in which $\mathcal{M}$ depends linearly on $\mathbf{B}$:

$$\mu_0 \mathcal{M} = \chi_B \mathbf{B}. \tag{28.2}$$

The *magnetic susceptibility* matrix χ_B is symmetric; this is a consequence of (28.1) and (28.2). In isotropic materials χ_B is a number, and in a linear magnet at rest $\mathbf{H} = \mathbf{B}/\mu_0 - \mathbf{M} = \mathbf{B}/\mu_0 - \mathcal{M}$ is related to $\mathbf{B}$ by

$$\mathbf{B} = \mu\mathbf{H}, \tag{28.3}$$

where

$$\mu = \frac{\mu_0}{1 - \chi_B} \tag{28.4}$$

is the *magnetic permeability*. Unlike the dielectric susceptibility, which is always positive, χ_B may be of either sign. Materials with positive χ_B are called *paramagnetic*; those with negative χ_B are called *diamagnetic*. Whether positive or negative, values of $|\chi_B|$ range from 10^{-9} to 10^{-4}. There is certainly no danger of the permeability according to (28.4) blowing up as $\chi_B \rightarrow 1$.

In the older tradition the magnetic susceptibility is defined differently, viz.:

$$\mathcal{M} = \chi_H \mathbf{H}. \tag{28.5}$$

In terms of χ_H, the permeability is

$$\mu = (1 + \chi_H)\mu_0. \tag{28.6}$$

The newer χ_B and the older χ_H are related by

$$(1 - \chi_B)(1 + \chi_H) = 1. \tag{28.7}$$

The two obviously have the same sign. In fact, they are equal to first order in either of them, and to distinguish between them would be a refinement that is not justified by the accuracy with which magnetic susceptibilities are measured.

The force on a linear magnetic material can easily be found by following the method that was used, in §23, for the analogous case of linear dielectrics. At a magnet–vacuum boundary this force may constitute a tension or a pressure, depending on whether the material is paramagnetic or diamagnetic. But its magnitude is very small, because magnetic susceptibilities are so small compared to electric susceptibilities.

A theory of elastic magnets can be developed by assuming various forms for the reduced free energy (cf. (25.4))

$$\varphi = \Phi(F^T F, \vartheta, F^T \mathbf{B}). \tag{28.8}$$

In particular, a simple expansion for weak magnetic fields leads to a description of permanent magnets, piezomagnets and linear magnets (including the isotropic dia- and paramagnetic materials). These are, of course, analogues of pyroelectrics, piezoelectrics and linear dielectrics.

The equations governing a steady magnetic field are

$$\mathbf{curl}\,\mathbf{H} = \mathbf{j}, \qquad \mathbf{n} \times [\![\mathbf{H}]\!] = 0,$$

$$\operatorname{div}\mathbf{B} = 0, \qquad \mathbf{n} \cdot [\![\mathbf{B}]\!] = 0,$$

$$\mathbf{H} = \mathbf{B}/\mu_0 - \boldsymbol{\mathcal{M}} \tag{28.9}$$

For permanent magnets these have non-trivial regular solutions, analogous to the electric fields of pyroelectrics, even in the absence of any currents.

PROBLEM

Problem 28–1. A spherical permanent magnet has uniform magnetization $\boldsymbol{\mathcal{M}}$ Find the fields $\mathbf{H}$ and $\mathbf{B}$ inside and outside the sphere.

For a linear – diamagnetic or paramagnetic – material, equations (28.9) have only trivial regular solutions when $\mathbf{j} = 0$. Such a material can therefore only act as an *electro-magnet*. If it fills all of the space around the circuits, the magnetic field is given by the formulae of the previous section, provided we replace μ_0 by μ everywhere. We shall dispense with a formulation of Thomson's theorem for magnetic fields. It can easily be done, and one of its corollaries is that, in linear magnetic media, the field lines of the solenoidal field $\mathbf{B}$ will be more concentrated in regions of high permeability.

PROBLEMS

Problem 28–2. A paramagnetic spherical shell with inner and outer radii a and b is placed in a uniform field $\mathbf{H}_0$. Show that the field $\mathbf{H}_i$ in the inner cavity

$r \leq a$ is uniform and parallel to $\mathbf{H}_0$, and that

$$\frac{H_i}{H_0} = \frac{9\mu\mu_0}{9\mu\mu_0 + 2(\mu - \mu_0)^2(1 - a^3/b^3)}.$$

Problem 28–3. A torus of soft iron with permeability μ has n turns of wire closely and evenly wound around it. A current i flows in the wire. Show that at points inside the torus $B = \mu ni/(2\pi r)$, where r is the distance from the axis of the torus. A small air-gap is formed in the iron by cutting away a thin sector bounded by two planes through the axis which make a small angle α with each other. Show that B in the gap is reduced from the foregoing value by the factor $1 + (\mu/\mu_0 - 1)\alpha/(2\pi)$.

Problem 28–4. A current i flows in a straight wire parallel to a semi-infinite block of material of permeability μ. Show that the magnetic field is given by an image system similar to that of Problem 22–5.

Problem 28–5. A current i flows in a straight wire parallel to a circular cylinder of permeability μ. Show that inside the cylinder $\mathbf{B}$ is $2\mu/(\mu + \mu_0)$ times as large as it would be if the cylinder were removed, and is everywhere in the same direction.

29. Ferromagnets

Ferromagnetic materials like iron, nickel and their alloys are the magnetic analogues of ferroelectric materials. Like the latter, they exhibit non-linear behaviour and hysteresis. They may be spontaneously magnetic, in a variety of domain structures, but only for temperatures below a characteristic temperature ϑ_c, called the (magnetic) *Curie temperature*. Above the Curie temperature, or *Curie point*, they are paramagnetic with a susceptibility χ_H that is inversely proportional to $\vartheta - \vartheta_c$; this is the (magnetic) *Curie-Weiss law*. The analogy between ferromagnetism and ferroelectricity is in fact so complete that their macroscopic treatment is governed by the same theory. There are, however, some *quantitative* differences, which we shall point out later.

If we compare the magnetostatic system of equations

$$\operatorname{div} \mu_0(\mathbf{H} + \boldsymbol{\mathcal{M}}) = 0, \qquad \mathbf{n} \cdot \mu_0 [\![\mathbf{H} + \boldsymbol{\mathcal{M}}]\!] = 0,$$

$$\operatorname{curl} \mathbf{H} = 0, \qquad \mathbf{n} \times [\![\mathbf{H}]\!] = 0, \tag{29.1}$$

with the electrostatic system (26.13), it is clear that the former is obtained from the latter by the replacements $\mathbf{E} \to \mathbf{H}$, $\epsilon_0 \to \mu_0$ and $\mathbf{P} \to \mu_0\,\mathcal{M}$. But this correspondence does not extend to the thermodynamic relations $\mathcal{M} = -\rho\varphi_{\mathbf{B}}$ and $\mathbf{P} = -\rho\varphi_{\mathcal{E}}$, because in $\varphi_{\mathbf{B}}$ the differentiation is with respect to $\mathbf{B}$. We shall therefore begin by replacing $\varphi(\rho, \vartheta, \mathbf{B})$ with another thermodynamic potential $\psi(\rho, \vartheta, \mathcal{G})$, where

$$\mathcal{G} = \mathbf{B}/\mu_0 - \mathcal{M} \qquad (29.2)$$

is the difference between the Galilei-invariant vectors $\mathbf{B}/\mu_0$ and $\mathcal{M}$. For a stationary body in an aether frame $\mathcal{G} = \mathcal{H}$ (cf. (10.9)), and either of these vectors equals the current potential $\mathbf{H}$ (cf. $(9.5)_1$ and $(10.6)_6$). In terms of $\mathcal{G}$, the relation $\rho\varphi_{\mathbf{B}} = -\mathcal{M}$ becomes

$$\rho\varphi_{\mathbf{B}}(\rho, \vartheta, \mathbf{B}) = \mathcal{G} - \mathbf{B}/\mu_0. \qquad (29.3)$$

We now regard (29.3) as an equation that (implicitly) determines $\mathbf{B}$ as a function of ρ, ϑ and $\mathcal{G}$, and then carry out a Legendre transformation to a new thermodynamic potential:

$$\rho\psi(\rho, \vartheta, \mathcal{G}) = \rho\varphi + (\mathbf{B} - \mu_0\mathcal{G})^2/(2\mu_0). \qquad (29.4)$$

Since $\mathbf{B}$ on the right hand side is now regarded as the solution of (29.3), the new thermodynamic potential ψ depends on ρ, ϑ and $\mathcal{G}$. Its $\mathcal{G}$-derivative is easily obtained by the chain rule:

$$\rho\psi_{\mathcal{G}}(\rho, \vartheta, \mathcal{G}) = -\mu_0\,\mathcal{M} \qquad (29.5)$$

The last result is analogous to $\rho\varphi_{\mathcal{E}} = -\mathbf{P}$. Clearly, then, we have replaced $\mathbf{B}$ by $\mathcal{G}$ as an independent variable. This replacement extends to the entropy inequality: taking the dot derivative of (29.4), we obtain

$$\rho\dot{\psi} = \dot{\rho}(\varphi - \psi) + \rho\dot{\varphi} + \mathcal{M} \cdot (\dot{\mathbf{B}} - \mu_0\dot{\mathcal{G}})$$

or

$$-\rho\dot{\varphi} - \mathcal{M} \cdot \dot{\mathbf{B}} = -\rho\dot{\psi} - \mu_0\,\mathcal{M} \cdot \dot{\mathcal{G}} + \tfrac{1}{2}\mu_0\mathcal{M}^2 I \cdot \operatorname{grad}\dot{\mathbf{x}}.$$

With the help of this relation, the entropy inequality (14.11) (with $\mathbf{P} = 0$) becomes

$$-\rho\dot{\psi} - \rho\eta\dot{\vartheta} - (\rho\mathbf{g} - \rho\dot{\mathbf{x}} - \epsilon_0\mathbf{E} \times \mathbf{B}) \cdot \ddot{\mathbf{x}} - \mu_0\,\mathcal{M} \cdot \dot{\mathcal{G}}$$

$$+ \left(\tau + \tfrac{1}{2}\mu_0\mathcal{M}^2 I\right) \cdot \operatorname{grad}\dot{\mathbf{x}} + \mathcal{J} \cdot \mathcal{E} - (\mathbf{q} \cdot \operatorname{grad}\vartheta)/\vartheta \geq 0. \qquad (29.6)$$

As in §14, we can now conclude that, if ϑ and $\dot{\mathbf{x}}$ are constant and uniform, and if, furthermore, $\mathcal{G}$ is constant and $\mathcal{E} = 0$, then ψ cannot increase. A state of a body for which ψ (assuming that it depends on further variables) is a minimum under these conditions will therefore be stable.

Equation (29.5) can be used to determine $\mathcal{G}$ as a function of ρ, ϑ and $\mathcal{M}$. Another Legendre transformation then leads to the thermodynamic potential (cf. (26.1))

$$\rho\tilde{\psi}(\rho, \vartheta, \mathcal{M}) = \rho\psi + \mu_0 \, \mathcal{M} \cdot \mathcal{G}. \tag{29.7}$$

This potential has the $\mathcal{M}$-derivative (cf. (26.2))

$$\rho\tilde{\psi}_{\mathcal{M}}(\rho, \vartheta, \mathcal{M}) = \mu_0\mathcal{G}. \tag{29.8}$$

Instead of specifying a magnetic material by a $\varphi(\rho, \vartheta, \mathbf{B})$ or by a $\psi(\rho, \vartheta, \mathcal{G})$, we can specify it by a $\tilde{\psi}(\rho, \vartheta, \mathcal{M})$. Consider now the thermodynamic potential (cf. (26.3))

$$\rho\tilde{\psi}(\rho, \vartheta, \mathcal{M}) = \rho\tilde{\psi}_0(\rho, \vartheta) + \mu_0[a(\vartheta - \vartheta_c)\mathcal{M}^2 + b\mathcal{M}^4], \tag{29.9}$$

where a and b are both positive and independent of ϑ or $\mathcal{M}$, and ϑ_c is a positive constant. Substitution of (29.9) in (29.8) gives (cf. (26.4))

$$[2a(\vartheta - \vartheta_c) + 4b\mathcal{M}^2] \, \mathcal{M} = \mathcal{G}. \tag{29.10}$$

We have already noted that, for a body at rest (in an aether frame), $\mathcal{G}$ is equal to the current potential $\mathbf{H}$. Thus

$$[2a(\vartheta - \vartheta_c) + 4b\mathcal{M}^2] \, \mathcal{M} = \mathbf{H} \tag{29.11}$$

is the equation that determines the relationship between $\mathcal{M}$ and $\mathbf{H}$ – the *magnetization curve*. The magnetization $\mathcal{M}$ is parallel or anti-parallel to $\mathbf{H}$, depending on the sign of the expression in the square brackets of (29.11). This is, of course, a consequence of the isotropy of the particular $\tilde{\psi}$ we have chosen.

If $\vartheta > \vartheta_c$, $\mathcal{M}$ is always parallel to $\mathbf{H}$. If $\mathbf{H}$ is small, the relationship is linear, with a susceptibility given by (cf. (26.5))

$$\chi_H = \frac{(2a)^{-1}}{\vartheta - \vartheta_c}. \tag{29.12}$$

This is the *Curie-Weiss law*. The critical temperature ϑ_c is the *Curie temperature* or the *Curie point*. For iron it is 1043 K. Below the Curie temperature the $\mathcal{M}(\mathbf{H})$ relationship is more complicated. For $\mathbf{H} = 0$ we have, besides $\mathcal{M} = 0$, the *spontaneous* ferromagnetic solution (cf. (26.7))

$$\mathcal{M}^2 = \frac{a(\vartheta_c - \vartheta)}{2b}. \tag{29.13}$$

Since it corresponds to the vanishing of the square bracket in (29.11), its direction is undetermined. (Again, this is a consequence of the isotropy of (29.9).) For $\mathbf{H} \neq 0$, every solution has a determinate direction, and we may assume that $\mathcal{M}$ and $\mathbf{H}$ both lie on the z axis:

$$[2a(\vartheta - \vartheta_c) + 4b\mathcal{M}_z^2]\mathcal{M}_z = H_z \tag{29.14}$$

(cf. (26.8)). The resulting magnetization curve is completely analogous to the ferroelectric polarization curve in Figure 7. Its stable and unstable branches are similarly determined by the minima and maxima of the thermodynamic potential (cf. (26.11))

$$\rho\psi = \rho\tilde{\psi}(\rho, \vartheta) + \mu_0[a(\vartheta - \vartheta_c)\mathcal{M}_z^2 + b\mathcal{M}_z^4 - \mathcal{M}_z H_z], \tag{29.15}$$

obtained by substituting (29.9) into (29.7). The investigation leads to the conclusion that, so long as $\mathbf{H}$ is not too large, there are two stable, oppositely directed, spontaneous magnetizations. Above the Curie temperature these ferromagnetic solutions disappear, and the material becomes paramagnetic.

Having now sufficiently pursued the analogy between ferroelectricity and ferromagnetism, we shall be content with a few general comments, especially as regards the (quantitative) *differences* between the two phenomena. Actual ferromagnets, like actual ferroelectrics, are crystalline or polycrystalline materials, and therefore have *magnetic* axes. These are, again, characterized by the relative adjectives 'easy' and 'hard'. Although *magnetostriction* certainly exists (it is responsible for the humming noise of transformers), it is generally a weaker effect, as compared with the non-magnetic stresses, than electrostriction is in ferroelectrics. We have noted, in §26, that single-domain ferroelectrics cannot exist (except in a conducting medium), because the depolarizing electric fields are much too strong. In ferromagnets the $\mathcal{M}(\mathbf{H})$ relation is usually such

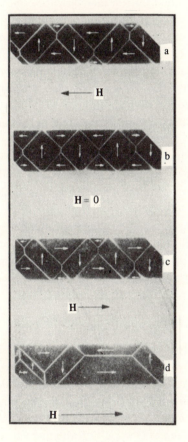

FIGURE 11

(From R. W. de Blois and C. D. Graham, *J. Appl. Phys.*, **29**, 931 (1959).)

that the demagnetizing (opposing) $\mathbf{H}$ that would result from (29.1) in a spontaneously, uniformly magnetized specimen is not strong enough to cause a reversal of $\mathcal{M}$. Single-domain ferromagnets are therefore possible, but are actually observed to occur only in very small specimens. Larger specimens have a magnetic domain structure. This is, again, explained as an effect of a surface free energy $\int \alpha \, dS$: although a single domain is not precluded by the magnetization curves of ferromagnets, subdivision into domains of finite size – which again turns out to vary as the square root $\sqrt{\ell}$ of the dimension of the body – corresponds to a minimum of the total free energy $\int \rho\psi \, dV + \int \alpha \, dS$. Only in ferromagnets that are smaller than a critical size, which depends on the surface tension coefficient α, does a single-domain configuration provide a minimum. In iron this critical size is about 20 nm.

If $\mathbf{H}$ and $\mathcal{M}$ are both uniform (the latter in each domain), the system (29.1) reduces to $\mathbf{n} \cdot [\![\mathcal{M}]\!] = 0$: the normal component of the magnetization must be continuous across the domain walls. Figure 11 shows some actual domain structures that were observed in a 50 μm iron whisker.

In a sufficiently strong applied field $\mathbf{H}$ the magnetization reaches saturation and becomes everywhere parallel to the field. This is, of course, a single-domain configuration that occurs in all ferromagnets, of any size. It does not contradict our former conclusions regarding single domains, because $- \mathcal{M} \cdot \mathbf{H}$ (the last term of (29.15)) is in this case *negative*; the minimum of the total free energy is therefore obtained with $\int \alpha \, dS = 0$.

30. Superconductors

In ordinary conductors (which we shall discuss fully in the next chapter) the total current i, which enters at one end of the conductor and leaves at another, is linearly related to the electric voltage drop V between the two ends. The ratio V/i depends on the material of which the conductor is made and on its shape, and is called the resistance of the conductor. Conduction of electricity through such conductors also leads to dissipation of energy, the rate of which is given by the volume integral of $\mathcal{J} \cdot \mathcal{E}$. This, as we shall prove in the next chapter, turns out to be proportional to the resistance.

In 1908 Kammerlingh Onnes succeeded in liquefying helium, which provided an ideal cold bath for experiments at temperatures within a few degrees from absolute zero. When he measured the resistances of various metals at these low temperatures, he made (in 1911) an intriguing discovery: some substances, when sufficiently cooled, conduct electricity without any resistance; they become *superconductors*. The currents that these superconductors transmit are of course accompanied by magnetic fields. These fields have a remarkable property: they do not enter the superconductor. In fact, when a material becomes superconducting, it expels any magnetic field that has previously existed inside it.† This magnetic field expulsion – called the *Meissner effect* – and the vanishing of resistance are not unconnected. In fact, the former implies the latter, because a magnetic field $\mathbf{B}$ that envelopes a body without penetrating is necessarily associated with surface currents $\mathbf{K} = \mathbf{n} \times \mathbf{B}/\mu_0$ (we assume the surroundings to be non-magnetic, so that $\mathbf{H} = \mathbf{B}/\mu_0$). Unlike the volume conduction currents in ordinary conductors, these surface currents do not require any voltage drop to drive them. Hence there is no resistance. Nor is there any energy dissipation, since $\int \boldsymbol{\mathcal{J}} \cdot \boldsymbol{\mathcal{E}} \, dV$ vanishes.

We can easily calculate the magnetic part of the stress on a superconductor. It is given by

$$T\mathbf{n} = [-B^2/(2\mu_0)I + \mathbf{B} \otimes \mathbf{B}/\mu_0]\mathbf{n}$$

$$= -B^2/(2\mu_0)\mathbf{n} + \mathbf{H}B_n = -B^2/(2\mu_0)\mathbf{n}, \tag{30.1}$$

since $B_n = 0$. This is a pressure force of $B^2/(2\mu_0) = \frac{1}{2}\mu_0 H^2$ per unit area.

PROBLEMS

Problem 30–1. A non-magnetic sphere of radius a is placed in a uniform field $\mathbf{B}_0$. Prove that when the sphere becomes superconducting, the magnetic field changes to

$$\mathbf{B} = \left(1 + \frac{a^3}{2r^3}\right)\mathbf{B}_0 - \frac{3a^3}{2r^3}(\mathbf{n} \cdot \mathbf{B}_0)\mathbf{n}, \tag{30.2}$$

---

† We are confining ourselves to type I superconductors. In type II superconductors the transition and the magnetic expulsion occur gradually.

where $\mathbf{n} = \mathbf{r}/r$. Show that the magnetic pressure is $9B_0^2 \sin^2 \theta/(8\mu_0)$, where θ is the spherical polar angle, measured from the direction of $\mathbf{B}_0$.

Problem 30–2. Prove that the magnetic stress on a superconductor is $\frac{1}{2}\mathbf{K} \times \mathbf{B}$, where $\mathbf{K} = \mathbf{n} \times \mathbf{B}/\mu_0$ is the surface current. This may be compared to the electrostatic stress $\frac{1}{2}\sigma\mathbf{E}$ on a conductor.

The Meissner effect, of which superconductivity is a direct result, is one reason for regarding superconductors as magnetic materials. There is another. When superconductors were discovered it was hoped that they would carry arbitrarily large (surface) currents without any resistance. That would have made them ideal electromagnets, capable of creating intense magnetic fields without dissipation of energy. These hopes were shattered with the discovery that a superconductor becomes normal when the (tangential) magnetic field exceeds a critical value $B_c = \mu_0 H_c$, which depends on the temperature. The typical dependence of H_c on ϑ, as established by experiments, is shown in Figure 12. The superconducting state is only possible for temperatures less than ϑ_c, and then only so long as the magnetic field is below the value given by the curve. For example, for tin the critical temperature is $\vartheta_c = 3.72\,\mathrm{K}$, and the largest possible magnetic field (which corresponds to zero absolute temperature) is $H_c = 25000$ amp/m.

The transition from the normal to the superconducting state has some further properties, which can be shown to follow from the principles laid down in Chapter IV. We assume that the superconducting and normal phases can both be described by the simple constitutive class (15.1) of fluids. Each of the two phases is then characterized by a free energy. For the normal phase, which we assume to be devoid of polarization or magnetization, we have (cf. (15.11)–(15.15) and (15.18)) the relations

$$\eta = -\varphi_\vartheta(\rho, \vartheta),$$

$$p = \rho^2 \varphi_\rho(\rho, \vartheta),$$

$$\mathbf{g} = \dot{\mathbf{x}} + \epsilon_0 \mathbf{E} \times \mathbf{B}/\rho,$$

$$T\mathbf{n} = -[p + \epsilon_0 E^2/2 + B^2/(2\mu_0)]\mathbf{n}$$

$$+\epsilon_0 E_n \mathbf{E} + B_n \mathbf{B}/\mu_0 + \epsilon_0 \mathbf{E} \times \mathbf{B}\dot{x}_n. \tag{30.3}$$

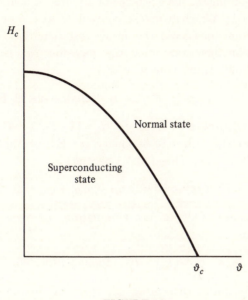

FIGURE 12

The relations for the superconducting phase are the same ones, except that the electromagnetic field vanishes. Of course the free energies of the two phases are not the same.

Consider now a surface separating a superconductor from its normal phase. We shall assume that the two phases have the same temperature ϑ. The other variables, namely the densities of the two phases and the (tangential) magnetic field, cannot then be arbitrarily specified, because they are subject to the jump conditions of §11–12. We must therefore determine the constraints that these conditions impose in the present case. The jump conditions (cf. (11.13), (12.3), (12.6) and (11.33)) are

$$[\![\rho(\dot{\mathbf{x}} - \mathbf{v})]\!] \cdot \mathbf{n} = 0, \qquad [\![\rho\mathbf{g} \otimes (\dot{\mathbf{x}} - \mathbf{v}) - T]\!]\mathbf{n} = 0,$$

$$[\![\rho\varepsilon(\dot{\mathbf{x}} - \mathbf{v}) - \dot{\mathbf{x}}T + \mathbf{q} + \boldsymbol{\mathcal{E}} \times \boldsymbol{\mathcal{H}}]\!] \cdot \mathbf{n} = 0,$$

$$[\![\rho\eta(\dot{\mathbf{x}} - \mathbf{v}) + \mathbf{q}/\vartheta]\!] \cdot \mathbf{n} \geq 0. \tag{30.4}$$

It should, perhaps, be noted that, with material crossing the magnetic field in the normal phase, the existence of an electric field on that side cannot be ignored. We shall find it convenient to choose a frame in which, at the instant considered, the surface is stationary, so that $v_n = 0$. The first condition then states that the mass flux $\rho \dot{x}_n$ is continuous. Turning next to $(30.4)_3$, we note that

$$\boldsymbol{\mathcal{E}} \times \boldsymbol{\mathcal{H}} \cdot \mathbf{n} - \dot{\mathbf{x}} \cdot T\mathbf{n} = \dot{x}_n[p - \epsilon_0 E^2/2 - B^2/(2\mu_0)] + \mathbf{n} \times \mathbf{E} \cdot \mathbf{B}/\mu_0. \quad (30.5)$$

This relation results from $\boldsymbol{\mathcal{E}} = \mathbf{E} + \dot{\mathbf{x}} \times \mathbf{B}$, $\boldsymbol{\mathcal{H}} = \mathbf{H} - \dot{\mathbf{x}} \times \mathbf{D} = \mathbf{B}/\mu_0 - \epsilon_0 \dot{\mathbf{x}} \times \mathbf{E}$ and $(30.3)_4$. Since the surface is stationary, $\mathbf{n} \times \mathbf{E}$ is continuous, so that it vanishes on either side. Thus $(30.4)_3$ becomes

$$[\![\rho \dot{x}_n[\varepsilon - \epsilon_0 E^2/(2\rho) - B^2/(2\mu_0\rho) + p/\rho] + q_n]\!] = 0. \quad (30.6)$$

Using the continuity of ϑ, we now eliminate q_n between $(30.4)_4$ and (30.6) and obtain

$$\rho \dot{x}_n[\![\varepsilon - \vartheta\eta - \epsilon_0 E^2/(2\rho) - B^2/(2\mu_0\rho) + p/\rho]\!] \leq 0. \quad (30.7)$$

This inequality states that mass can only flow in the direction in which the expression in the double square brackets decreases. If we further assume that the mass flux depends continuously on ρ, ϑ, $\mathbf{E}$ and $\mathbf{B}$, it follows that the mass flux and the jump in (30.7) must vanish together (cf. the derivation of (16.14)). In equilibrium, when $\dot{x}_n = 0$, the expression in the double square brackets of (30.7) will therefore have the same value for both phases. Recalling the definition (14.10) of the free energy, we can express this condition for equilibrium in the form

$$\varphi_s(\rho_s, \vartheta) + \frac{p_s(\rho_s, \vartheta)}{\rho_s} = \varphi_n(\rho_n, \vartheta) + \frac{p_n(\rho_n, \vartheta)}{\rho_n}. \quad (30.8)$$

This is a relation between the densities of the phases and the temperature.

The function

$$\gamma = \varphi(\rho, \vartheta) + \frac{\rho^2 \varphi_\rho(\rho, \vartheta)}{\rho} \quad (30.9)$$

is called the (specific) *Gibbs thermodynamic potential*. If we regard the equation $p = \rho^2 \varphi_\rho(\rho, \vartheta)$ as defining the density ρ in terms of p and ϑ, the Gibbs potential becomes a function of p and ϑ. The equilibrium condition (30.8) is then

$$\gamma_s(p_s, \vartheta) = \gamma_n(p_n, \vartheta). \tag{30.10}$$

PROBLEM

Problem 30–3. Prove that $\gamma(p, \vartheta)$ has the partial derivatives

$$\frac{\partial \gamma(p, \vartheta)}{\partial \vartheta} = -\eta, \qquad \frac{\partial \gamma(p, \vartheta)}{\partial p} = \frac{1}{\rho}. \tag{30.11}$$

We have not yet made any use of the jump condition (30.4)$_2$. In equilibrium it requires $T\mathbf{n}$ to be continuous. According to (30.3)$_4$ this means

$$p_s(\rho_s, \vartheta) = p_n(\rho_n, \vartheta) + B^2/(2\mu_0), \tag{30.12}$$

which is a second relation between the densities of the phases. It states that the density of the superconductor will have a value corresponding to pressure equilibrium, not with p_n but with $p_n + B^2/(2\mu_0)$. Of course the extra pressure is just the one we have found in (30.1).

In actual experimental situations the normal state does not extend to infinity. It is usually bounded, and is therefore itself in pressure equilibrium with another material (such as the atmosphere). It is thus more practical to regard p_n, rather than ρ_n, as given (along with the temperature ϑ). We denote this pressure simply by p (e.g., one atmosphere) and combine the equilibrium conditions (30.10) and (30.12) in the form

$$\gamma_s(p + \tfrac{1}{2}\mu_0 H^2, \vartheta) = \gamma_n(p, \vartheta); \tag{30.13}$$

it is customary to use H rather than B/μ_0 in treatments of superconductivity. According to this equation, equilibrium of a superconductor with its normal state at a pressure p and temperature ϑ requires that the magnetic field have a definite, or *critical*, value $H_c(p, \vartheta)$. Since, by (30.11)$_2$, γ_s has a positive derivative with respect to its first argument, a higher field will lead to $\gamma_s > \gamma_n$. That will result in a transition to the normal state. Similarly, $H < H_c(p, \vartheta)$ will result in a transition from the normal to the superconducting state. When $H = 0$ the transition, at given p, will take place at a definite critical transition temperature

ϑ_c. These considerations are in accord with Figure 12, which shows the critical field H_c as a function of temperature for a given 'external' pressure p.

Let us now differentiate (30.13) with respect to the temperature, keeping p fixed. Using (30.11), we obtain

$$\eta_n - \eta_s = -\frac{1}{\rho_s}\frac{\partial}{\partial\vartheta}[\tfrac{1}{2}\mu_0 H_c^2(p,\vartheta)]. \qquad (30.14)$$

In order to interpret this result, we turn back to the jump condition (30.6), which up to this point has only been used in conjunction with $(30.4)_4$. According to (30.6), mass flow with a discontinuity (of magnitude $\mathcal{L}$, say) in the expression multiplying $\rho\dot{x}_n$ is associated with a discontinuity of the heat flux q_n at the surface of separation. The discontinuity $\mathcal{L}$ is therefore called the (specific) *latent heat of transition*. Now the jumps by which the continuous $\rho\dot{x}_n$ is multiplied in (30.6) and (30.7) differ exactly by $[\![\vartheta\eta]\!] = \vartheta[\![\eta]\!]$. In the limit of equilibrium the jump in (30.7) vanishes, so that the latent heat $\mathcal{L}$ (per unit mass) becomes ϑ times the equilibrium jump of η. From (30.14) we therefore obtain

$$\mathcal{L} = -\frac{\vartheta}{\rho_s}\frac{\partial}{\partial\vartheta}[\tfrac{1}{2}\mu_0 H_c^2(p,\vartheta)] \qquad (30.15)$$

for the equilibrium limit of the latent heat (per unit mass) of transition from the superconducting to the normal state. According to the typical Figure 12, $\mathcal{L}$ is positive, since the slope of H_c with ϑ is negative. The latent heat vanishes at $\vartheta = 0$ and at $\vartheta = \vartheta_c$ (where $H_c = 0$). These results, too, are confirmed by experiments on superconductors.

The specific heat at constant pressure is defined by $c = \vartheta\partial\eta(p,\vartheta)/\partial\vartheta$. We can calculate the difference between the specific heats of the two phases at equilibrium by differentiating (30.14) at constant p. It must, of course, be remembered that η_s and ρ_s are the partial derivatives of the γ_s of (30.13) and therefore have the same arguments as the latter; and that, furthermore, $\partial\eta(p,\vartheta)/\partial p = \rho^{-2}\partial\rho(p,\vartheta)/\partial\vartheta$ (which follows from (30.11)). The result of this calculation is

$$c_n - c_s = -\frac{\vartheta}{\rho_s}\frac{\partial^2}{\partial\vartheta^2}\tfrac{1}{2}\mu_0 H_c^2 - \frac{2\mathcal{L}}{\rho_s}\left(\frac{\partial\rho_s}{\partial\vartheta} + \frac{\partial\rho_s}{\partial p}\frac{\partial}{\partial\vartheta}\tfrac{1}{2}\mu_0 H_c^2\right). \qquad (30.16)$$

At $\vartheta = \vartheta_c$ both H_c and $\mathcal{L}$ vanish. Then

$$c_n - c_s = -\frac{\mu_0 \vartheta_c}{\rho_s} \left[\frac{\partial H_c(p, \vartheta_c)}{\partial \vartheta} \right]^2 . \tag{30.17}$$

A pronounced jump in the specific heat at the transition temperature had, in fact, been observed before the thermodynamic theory of super-conductors was developed. Equation (30.17), which is known as Rut-gers's formula, relates this jump to the transition temperature, the density and the slope of H_c at $\vartheta = \vartheta_c$, all of which are measurable. Rutgers's formula has been verified by numerous experiments.

PROBLEMS

Problem 30–4. Show, by differentiating (30.13) with respect to p, that

$$\frac{1}{\rho_n} - \frac{1}{\rho_s} = \frac{\partial}{\partial p} [\tfrac{1}{2} \mu_0 H_c^2(p, \vartheta)]. \tag{30.18}$$

This difference between the specific volumes of the phases at equilibrium vanishes (with H_c) at $\vartheta = \vartheta_c$.

Problem 30–5. Show, by differentiating (30.13) with respect to ϑ at fixed H_c, that

$$\frac{\partial p(\vartheta, H_c)}{\partial \vartheta} = \frac{\mathcal{L}}{\vartheta(\rho_n^{-1} - \rho_s^{-1})}. \tag{30.19}$$

This is the rate of change, with respect to temperature, of the pressure needed to keep the magnetic field critical.

Consider now a spherical superconductor in a magnetic field which is uniform at infinity (Problem 30–1). Near the sphere the enveloping magnetic field is non-uniform and has its greatest intensity at the 'equator'. If the field is gradually increased, it will at first exceed the critical value near the equator. It might be supposed that this will cause the material in an equatorial belt to become normal, the rest of the sphere remaining in the superconducting state; and that the boundary between the two phases will gradually move inward as the external magnetic field is in-creased, until the whole sphere becomes normal (and the field uniform). Throughout this process the field must be critical at every point on the phase boundary; it must also increase as one moves from the boundary into the normal belt, for otherwise a part of the normal region will again become superconducting. It turns out that these requirements cannot be met with any configuration in which the superconducting region is simply connected. What happens then is a breakup of the whole sphere into a complicated structure of normal and superconducting domains, called a *mixed state*. Such structures have in fact been observed. As with ferroelectrics and ferromagnets, a surface free energy is invoked in order to account for the eventual halting of the breakup, and for its variation between different materials.

PROBLEM

Problem 30–6. A small permanent magnet of moment **m** is placed near a super-conductor. Show that the magnetic field is obtained by adding the field of an image magnet **m'** inside the superconductor. Deduce from the known behaviour of a pair of magnets that **m** is repelled by the superconductor.

CHAPTER IX

Conductors

31. Linearly conducting materials

We have shown, in our discussion of (non-viscous) fluids and elastic materials, that the conduction current $\mathcal{J}$ and the heat flux vector $\mathbf{q}$ must both vanish whenever $\mathcal{E}$ and $\operatorname{grad}\vartheta$ are both zero. For small electromotive intensities and temperature gradients we were then led to the concept of linearly conducting materials. In these materials we have the homogeneous relations

$$\mathcal{J} = a\mathcal{E} + b\operatorname{grad}\vartheta,$$

$$\mathbf{q} = c\mathcal{E} + d\operatorname{grad}\vartheta, \tag{31.1}$$

where a, b, c and d are matrices with elements depending on F, ϑ and $\mathbf{B}$ in elastic materials, and on ρ, ϑ and $\mathbf{B}$ in fluids (cf. (16.15)). Writing $a = \frac{1}{2}(a + a^T) + \frac{1}{2}(a - a^T)$ and introducing a vector $\mathbf{a}$ through the equation (cf. $(4.2)_2$)

$$\epsilon_{ijk}a_k = \tfrac{1}{2}(a_{ij} - a_{ji}),$$

we have

$$a\mathcal{E} = \tfrac{1}{2}(a + a^T)\mathcal{E} + \mathcal{E} \times \mathbf{a}. \tag{31.2}$$

Instead of (31.1), we can therefore write for linearly conducting materials the relations

$$\mathcal{J} = a\mathcal{E} + \mathcal{E} \times \mathbf{a} + b\operatorname{grad}\vartheta + \operatorname{grad}\vartheta \times \mathbf{b},$$

$$\mathbf{q} = c\mathcal{E} + \mathcal{E} \times \mathbf{c} + d\operatorname{grad}\vartheta + \operatorname{grad}\vartheta \times \mathbf{d}, \tag{31.3}$$

143

where a, b, c, and d are now *symmetric* matrices. The symmetric matrices a, ..., d and the vectors $\mathbf{a}$, ..., $\mathbf{d}$ express material properties. They are not all independent, because the inequality (16.7) imposes restrictions on them. For example, the symmetric matrices a and $(-d)$ must be positive-definite. For highly isotropic materials in weak magnetic fields each of the matrices reduces to a multiple of the identity, and each of the vectors reduces to a multiple of $\mathbf{B}$. For such materials, the relations (31.3) are commonly written in the form

$$\mathcal{E} = \mathcal{J}/\sigma + \alpha \operatorname{\mathbf{grad}} \vartheta + \mathcal{R}\mathbf{B} \times \mathcal{J} + \mathcal{N}\mathbf{B} \times \operatorname{\mathbf{grad}} \vartheta,$$

$$\mathbf{q} = \Pi \mathcal{J} - \kappa \operatorname{\mathbf{grad}} \vartheta + \mathcal{S}\mathbf{B} \times \mathcal{J} + \mathcal{L}\mathbf{B} \times \operatorname{\mathbf{grad}} \vartheta. \tag{31.4}$$

Each term in (31.4) is responsible for a well known effect. For example, the term $\mathcal{R}\mathbf{B} \times \mathcal{J}$ in $(31.4)_1$ gives rise to an electromotive intensity which is perpendicular to the conduction current and to the magnetic field; this is called the *Hall effect*. The term with the coefficient $\mathcal{N}$ is connected to an $\mathcal{E}$ which is normal to both the magnetic field and the temperature gradient; this is the *Nernst effect*. The third term on the right hand side of $(31.4)_2$ gives rise to a heat flux perpendicular to the magnetic field and to the conduction current; this is the *Ettingshausen effect*. The last term, with the coefficient $\mathcal{L}$, gives a heat flux normal to the temperature gradient and to the magnetic field; this is called the *Leduc-Righi effect*.

If the transverse terms are absent, as they must be whenever the magnetic field is negligible, we have

$$\mathcal{E} = \mathcal{J}/\sigma + \alpha \operatorname{\mathbf{grad}} \vartheta,$$

$$\mathbf{q} = \Pi \mathcal{J} - \kappa \operatorname{\mathbf{grad}} \vartheta. \tag{31.5}$$

If α, called *the Thomson coefficient*, vanishes, $(31.5)_1$ becomes Ohm's law $\mathcal{J} = \sigma \mathcal{E}$. The coefficient σ is called *the electric conductivity*. If Π, called *the Peltier coefficient*, vanishes, $(31.5)_2$ reduces to Fourier's law $\mathbf{q} = -\kappa \operatorname{\mathbf{grad}} \vartheta$. The coefficient κ is called *the thermal conductivity*. It is easy to obtain the conditions that the coefficients σ, α, Π and κ must satisfy, because, with (31.5), (16.7) becomes an inequality for a quadratic form. The conditions are:

$$\sigma \geq 0, \qquad \kappa \geq 0,$$

$$4(\vartheta\kappa/\sigma) \geq (\Pi - \vartheta\alpha)^2. \tag{31.6}$$

32. Charge relaxation

We have seen that, in linearly conducting materials of the simplest kind, the electromotive intensity, the conduction current density and the temperature gradient are related by

$$\mathcal{E} = \mathcal{J}/\sigma + \alpha \operatorname{\mathbf{grad}} \vartheta. \tag{32.1}$$

The electric conductivity σ has the dimensions of current density divided by electric field, or amp/(volt·m). The ratio volt/amp is called the *ohm* and is the SI unit of resistance (which we shall define in the next section). The reciprocal of the ohm is called the *siemen*.† Thus σ has the SI units of $(ohm·m)^{-1}$ or siemen/m. Metallic conductors – like copper, silver or zinc – have conductivities of a few 10^7 siemen/m.

If the second term on the right hand side of (32.1) can be neglected (this will be the case when the current is large, or the temperature uniform), we have Ohm's law,

$$\mathcal{J} = \sigma\mathcal{E}. \tag{32.2}$$

For a material at rest, this takes the form $\mathbf{j} = \sigma\mathbf{E}$. If we substitute this in the law of charge conservation, we obtain

$$q_t + \operatorname{div}\mathbf{j} = q_t + \operatorname{div}\sigma\mathbf{E},$$

$$= q_t + \sigma\operatorname{div}\mathbf{E} + \mathbf{E}\cdot\operatorname{\mathbf{grad}}\sigma = 0. \tag{32.3}$$

If the conductivity and the permittivity are uniform, $q = \operatorname{div}\mathbf{D} = \epsilon\operatorname{div}\mathbf{E}$ gives

$$q_t + \frac{\sigma}{\epsilon}q = 0. \tag{32.4}$$

We conclude that the charge density q decays as $e^{-(\sigma/\epsilon)t}$ wherever (32.4) holds. The same is true of $\operatorname{div}\mathbf{j} = -q_t$. For metallic conductors, with σ

† In the older literature it was called *mho* (the word *ohm* read backwards).

of the order of several times 10^7 siemen/m, ϵ/σ is of the order of 10^{-18} seconds, and any q will disappear practically instantaneously. Since charge is conserved, it cannot really disappear, and we may rightly ask: where does it go? The answer is that it accumulates at those places where (32.4) does *not* hold; in other words, on the surfaces along which σ or ϵ are discontinuous. Electrostatics is therefore only a special example of the rule that the electric charge of a conductor must reside wholly on its surface. Except for phenomena involving frequencies $\omega \geq \sigma/\epsilon$ – which are far in excess of optical frequencies – we may disregard this rapid process of charge relaxation in any conductor. In fact, if we want to avoid getting bogged down with what happens during the first 10^{-18} seconds, we *must* disregard the law of charge conservation and replace it by div $\mathbf{j} = 0$.

Noting that $D_t/j = \epsilon E_t/j$, too, is comparable to $\omega\epsilon/\sigma$, so that the displacement term may be dropped from Maxwell's equations, we conclude that Ohmic conductors (once charge relaxation has taken place) are governed by the equations

$$\mathbf{curl\ H} = \mathbf{j},$$

$$\mathbf{j} = \sigma\mathbf{E},$$

$$\mathrm{div\ } \mathbf{B} = 0,$$

$$\mathbf{curl\ E} = -\mathbf{B}_t. \tag{32.5}$$

The first of these equations already ensures that div $\mathbf{j} = 0$. The system (32.5) must be supplemented by information regarding the magnetization $\mathbf{M} = \mathbf{B}/\mu_0 - \mathbf{H}$.

It is important to note that the system (32.5) holds in the interior, but not on the surface, of a conductor. For example, if the plates of a charged capacitor are connected by a conducting wire, the ensuing discharge may be calculated by applying (32.5) to the wire, but the surface charges $\pm Q(t)$ on the plates cannot be ignored along with the charge density q in the wire. On the contrary, the law of charge conservation, in the form $i(t) = dQ/dt$, is essential for calculating the discharge.

For non-magnetizable and non-polarizable conductors we have shown in §15–16 that the free energy φ is independent of $\mathcal{E}$ or $\mathbf{B}$. So is the entropy $\eta = -\varphi_\vartheta$. We have also shown that the law of energy balance takes the form

$$\rho\vartheta\dot{\eta} = \boldsymbol{\mathcal{J}} \cdot \boldsymbol{\mathcal{E}} + \rho h - \operatorname{div} \mathbf{q}; \tag{32.6}$$

(cf. (15.23)). If we substitute $\boldsymbol{\mathcal{J}} = \mathbf{j} = \sigma\boldsymbol{\mathcal{E}}$, this becomes

$$\rho\vartheta\dot{\eta} = j^2/\sigma + \rho h - \operatorname{div} \mathbf{q}. \tag{32.7}$$

The last two terms on the right hand side of this equation are, by definition, the heating per unit volume (cf. (11.23)–(11.28)). So far as changes in the entropy are concerned, the added term j^2/σ acts just like positive heating, since $\sigma \geq 0$. For example, if the dependence of η on ϑ (the specific heat) is such that positive heating causes the body to warm up (increases its temperature), then j^2/σ will have the same effect. It is for this reason that j^2/σ is, somewhat confusingly, called *Joule heating* (per unit volume). Actually, it is neither heating nor working; the origin of this term lies in the *extra* energy flux $\boldsymbol{\mathcal{E}} \times \boldsymbol{\mathcal{H}}$ we have introduced in the basic law of energy balance (cf. (12.5)). In matters of balance, as with money, one man's loss is another man's gain. The Joule heating term j^2/σ is therefore also referred to as *Ohmic loss*. Perhaps *Joule warming* would be a much better name than *Joule heating*, referring, correctly, to what it does, rather than, incorrectly, to what it is. This warming effect is responsible for the operation of all electrical heating elements and fuses.

PROBLEM

Problem 32–1. A large sphere of radius a is made of material of conductivity σ and permittivity ϵ. At $t = 0$ a charge Q_0 is uniformly distributed over the surface of a small concentric sphere $r = a$. Determine how the charge Q on the inner sphere varies with time. If the temperature is initially uniform and the sphere has a heat capacity C, find the ultimate change in temperature.

33. Resistance

Consider now the steady flow of electric current through a stationary, conducting medium between two conductors, which we shall call the *electrodes*. The electrodes may, for example, be the two ends – or cross sections – of a conducting wire, the inner and outer surfaces of a hollow, conducting, cylinder, etc. According to (32.5), the following equations hold in the medium:

$$\text{div}\,\mathbf{j} = 0,$$

$$\mathbf{j} = \sigma\mathbf{E},$$

$$\text{curl}\,\mathbf{E} = 0. \tag{33.1}$$

The last of these can be satisfied, as usual, by introducing an electric potential. As for the electrodes, we assume that the current leaves or enters them normally. According to $(33.1)_2$ we shall then have, on each electrode,

$$\mathbf{n} \times \mathbf{E} = 0. \tag{33.2}$$

The requirement that the same current i leaves one electrode and enters another is expressed by

$$\int j_n\,dS = \pm i, \tag{33.3}$$

with the integral taken over the surface of the electrode.

Having found the distribution of $\mathbf{j}$ and $\mathbf{E}$ (or the electric potential) throughout the medium, we define its *resistance* R as

$$R = \frac{V}{i}, \tag{33.4}$$

with the signs of the voltage $V = \int E_s\,ds$ between the electrodes, and of the current $i = \int j_n\,dS$ leaving or entering an electrode, chosen in such a way as to render R positive.

It is easy to see that the calculation of resistance requires the solution of a potential problem. But it is even more profitable to compare it with the calculation of the capacity of a dielectric capacitor. The latter problem, it will be recalled, is to solve the equations

$$\text{div}\,\mathbf{D} = 0,$$

$$\mathbf{D} = \epsilon\mathbf{E},$$

$$\text{curl}\,\mathbf{E} = 0. \tag{33.5}$$

These must be solved with the conditions that

$$\mathbf{n} \times \mathbf{E} = 0 \tag{33.6}$$

on each of the plates, and that the charges on the plates are

$$\int D_n \, dS = \pm Q. \tag{33.7}$$

The capacity is then defined by

$$C = \frac{Q}{V}. \tag{33.8}$$

If the conductivity σ of the resistor and the permittivity of the capacitor are distributed in the same way, the two problems are not merely similar; they are, mathematically speaking, the same problem. In particular, if σ and ϵ are constants, we have the simple relation

$$RC = \frac{\epsilon}{\sigma}. \tag{33.9}$$

Thus a parallel-plate resistor of cross section S and length d has a resistance of

$$R = \frac{d}{\sigma S} \tag{33.10}$$

(cf. (19.21)). A coaxial resistor – a hollow cylinder with inner radius a, outer radius b and length L – has a resistance of (cf. (22.11))

$$R = \frac{\ln(b/a)}{2\pi\sigma L}. \tag{33.11}$$

PROBLEM

Problem 33–1. A spherical resistor consists of a hollow sphere with inner radius a and outer radius b, made of material with conductivity σ. Calculate its resistance. Use the answer to determine the capacity of a spherical capacitor which is filled with a dielectric of permittivity ϵ.

At a junction formed by several conductors, div $\mathbf{j} = 0$ requires that the sum of the outgoing currents equal the sum of the ingoing ones. This is *Kirchhoff's first rule*:

$$\sum i = 0. \tag{33.12}$$

The currents i in this sum are to be taken with their 'correct algebraic signs'; this just means that all outgoing currents must be given one sign, and all ingoing currents, the opposite sign.

The entire Joule warming in the medium between the electrodes is

$$\int \mathbf{j} \cdot \mathbf{E} \, d^3 x = - \int \mathbf{j} \cdot \mathbf{grad}\, V \, d^3 x = - \int \operatorname{div}(V\mathbf{j}) \, d^3 x. \qquad (33.13)$$

The last term can be transformed into an integral of $V j_n$ over the surface of the medium. Noting that the potential is constant on each electrode, that j_n vanishes at the boundary between the conducting medium and the insulating surroundings, and that the total current through the medium is given by (33.3), we find that the Joule warming can be simply expressed, in terms of the total current and the medium's resistance, as $i^2 R$.

The mathematical equivalence between the problems of determining capacity and resistance extends, of course, to Thomson's theorem of §22. For any solenoidal distribution of the current density $\mathbf{j}$ – a distribution satisfying $(33.1)_1$, but not necessarily $(33.1)_{2,3}$ – with given total currents $\oint j_n \, dS$ issuing from, or entering into, a finite system of electrodes, the Joule warming,

$$W[\mathbf{j}] = \int \frac{j^2}{\sigma} \, d^3 x, \qquad (33.14)$$

may be regarded as a functional of $\mathbf{j}$. By Thomson's theorem (§22), W will be minimal when $\mathbf{j} = \sigma \, \mathbf{grad}\, V$, with V constant on each electrode. In particular, j^2 will be large in regions of high conductivity.

The foregoing variational formulation of the potential (or resistance) problem for Ohmic conductors does not correspond to the usual method of passing a current through a conducting medium. Rather than prescribe the total currents through the electrodes, we usually maintain each one at a given *potential*. Consider then the Joule warming in the form

$$W[V] = \int \sigma (\mathbf{grad}\, V)^2 \, d^3 x. \qquad (33.15)$$

For a given distribution of $\sigma(\mathbf{x}) \geq 0$, we regard $W[V]$ as a functional of $V(\mathbf{x})$, the latter being subject only to the requirement of having prescribed constant values on the electrodes. We do *not* require V to satisfy

$\mathrm{div}(\sigma\,\mathbf{grad}\,V) = 0$. Clearly, then, $W[V]$ will have no maximum, for we can let V oscillate wildly between the electrodes. It will, however, have a minimum – a positive one, unless the electrodes are all maintained at the same potential. In order to find it, we calculate the variation of $W[V]$. Taking account of the fact that $\delta V = 0$ on the electrodes (where V is prescribed), integration by parts gives

$$\delta W = -2 \int \delta V \, \mathrm{div}(\sigma\,\mathbf{grad}\,V)\, d^3 x. \qquad (33.16)$$

Hence the minimum is obtained when V satisfies $\mathrm{div}(\sigma\,\mathbf{grad}\,V) = 0$. This V is precisely the solution of the potential problem (33.1). The result (sometimes referred to as The Principle of Minimum Dissipation) tells us that the actual distribution of electric potential throughout a conducting medium, in which a finite system of electrodes with pre-scribed potentials is embedded, is always the one for which the entire Joule warming is least. Every electric toaster, for example, works in the worst possible way (a feature which is not usually advertised, and for which the manufacturer can hardly be blamed). Another conse-quence of this variational formulation is that the lines of electric field $\mathbf{E} = -\mathbf{grad}\,V$ are more concentrated in regions of *low* conductivity.

PROBLEMS

Problem 33–2. A spherical hole of radius a is cut out of an infinite block of uniform conductivity. At large distances the current is uniform and in the z direction. Show that $V = A[r + a^3/(2r^2)]\cos\theta$ and that the lines of current flow are given by $(r^3 - a^3)\sin^2\theta = \mathrm{const} \times r$.

Problem 33–3. Prove that in a system of electrodes with zero total net current in an Ohmic medium, at least one electrode has current leaving everywhere, and one has current entering everywhere.

Problem 33–4. Prove that Green's reciprocal theorem $\sum iV' = \sum i'V$ holds for steady currents between electrodes in an Ohmic medium.

Problem 33–5. A closed curve C in the xy plane, lying entirely on the positive side of the y axis, is rotated through $180°$ about the y axis. The volume so formed is filled with a material of uniform conductivity σ, and electrodes are placed at the two plane ends. Prove that the resistance R is given by

$$\frac{1}{R} = \frac{\sigma}{\pi} \int \frac{dS}{x},$$

where the integration is over the area enclosed by C.

Problem 33–6. A thin spherical shell of radius a and thickness t is made of material with conductivity σ. Current enters and leaves by two small spherical electrodes of radius c whose centres are at the ends A, B of a diameter AOB. If i is the total current and P is a point on the shell such that the angle POA is θ, show that the current density at P is $i/(2\pi\sigma t \sin\theta)$, and that the resistance of the shell is

$$\frac{1}{\pi\sigma t} \ln \cot \frac{c}{2a}.$$

Problem 33–7. The space between two parallel conducting planes is filled with two slabs, one of thickness a_1, conductivity σ_1 and permittivity ϵ_1, and the other of thickness a_2, conductivity σ_2 and permittivity ϵ_2. Find the current and the surface charge density on the boundary between the slabs when a voltage V is applied to the conducting planes.

34. Electromotive force

The foregoing considerations were based on Ohm's law $\mathbf{j} = \sigma\mathbf{E}$, which is a special case of the more general relation (32.1). If we integrate (32.1) along a line, we obtain

$$\int \mathcal{E}_s \, ds = \int \frac{J_s}{\sigma} \, ds + \int \alpha \frac{\partial \vartheta}{\partial s} \, ds. \tag{34.1}$$

The line integral of the second, Thomson term on the right hand side, taken with the *opposite sign*, is called the *Thomson electromotive force*, usually abbreviated to *Thomson emf*, and denoted $\mathcal{E}_T$. It is, of course, not a force; its units are those of electric field times distance, that is, volts. The notation is standard; the scalar $\mathcal{E}$ should not be confused with the electromotive intensity *vector* $\boldsymbol{\mathcal{E}}$. If we apply (34.1) to linear currents – which flow along wires, idealized as lines – the first term on the right hand side becomes the product of current by resistance:

$$\int \mathcal{E}_s \, ds = Ri - \mathcal{E}_T. \tag{34.2}$$

For a stationary conductor in a steady magnetic field, the integral on the left hand side is independent of the path of integration, since $\mathcal{E} = \mathbf{E}$ and $\mathbf{curl}\ \mathbf{E} = 0$; it is simply the voltage between the ends. In this case we therefore have

$$V_1 - V_2 = Ri - \mathcal{E}_T, \tag{34.3}$$

where R is the resistance between the points 1 and 2 along the current line.

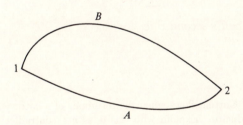

FIGURE 13

A *thermocouple* is obtained by joining two different conductors as shown in Figure 13. If the Thomson coefficients are constants, each Thomson emf is just $-\alpha \int d\vartheta$, and if we take the integrals in (34.2) around the whole loop (or apply (34.3) to each conductor), we obtain

$$(R_A + R_B)i + (\alpha_A - \alpha_B)(\vartheta_2 - \vartheta_1) = 0, \tag{34.4}$$

where i is counted as positive if it flows counterclockwise. Thus a current will flow if the conductors are different ($\alpha_A \neq \alpha_B$) and the junctions are at different temperatures. This arrangement can be used for measuring any one of the quantities that appear in (34.4) – resistance, current, Thomson coefficient, temperature – if all the other quantities are known. Thomson coefficients are of the order of microvolts per Kelvin degree. They lead to rather small currents, of the order of milliamperes. This means that the Thomson emf can be neglected whenever the current is large.

The Thomson emf follows, basically, from the constitutive assumptions we made in §15–16. In a battery, or voltaic cell, there are inhomogeneous chemical concentrations. A constitutive theory of such cells can be constructed by including concentration gradients among the arguments of the constitutive relations. This would lead to a concentration, or *battery*, emf $\mathcal{E}$, analogous to the Thomson emf. For a closed current loop, consisting of several elements, each with a resistance R, current i and emf $\mathcal{E}$, summation of relations like (34.3) gives

$$\sum (Ri - \mathcal{E}) = 0. \tag{34.5}$$

This is *Kirchhoff's second rule*.

PROBLEM

Problem 34–1. Current flows when the terminals of a battery are connected by an Ohmic conductor. Explain why this does not contradict the statement of Problem 33–3.

35. The skin effect

A stationary Ohmic conductor is governed by the equations (32.5). With $\mathbf{j}$ eliminated, these take the form

$$\mathbf{curl}\ \mathbf{H} = \sigma \mathbf{E},$$

$$\mathrm{div}\ \mathbf{B} = 0,$$

$$\mathbf{curl}\ \mathbf{E} = -\mathbf{B}_t. \tag{35.1}$$

We have already discussed solutions of this system for steady fields, when $\mathbf{B}_t = 0$. Now we wish to consider variable fields that are still sufficiently weak for the conduction current to be linear in the electromotive intensity. This introduces a complication, because $\mathcal{J}$ may fail to keep up with $\mathcal{E}$ when the latter changes rapidly. We shall therefore replace $\mathcal{J} = \sigma\mathcal{E}$ by a more general, history-dependent, linear relation of the form

$$\mathcal{J}(t) = \int_0^\infty f(\tau)\mathcal{E}(t - \tau)\,d\tau, \tag{35.2}$$

where $f(\tau)$ must, of course, be such that the integral will converge. This is the same as saying that electromotive intensities in the distant past must have a vanishing effect. Without an assumption of this kind, experimental physics would become impossible, for the physicist is usually ignorant of the *complete* history of any material he uses. He therefore works on the assumption that, if a material has any memory of the past, it is a *fading memory*.

If we substitute the right hand side of (35.2) (with $\mathcal{E} = \mathbf{E}$ for a stationary conductor) for $\sigma\mathbf{E}$, the system (35.1) remains linear, but it ceases to be a system of partial differential equations. We therefore resolve the fields by a Fourier expansion into *monochromatic* components, each of which depends on the time through the factor $e^{-i\omega t}$. For such monochromatic fields, the history-dependent Ohm's law (35.2) reduces to $\mathcal{J} = \sigma(\omega)\mathcal{E}$, where the frequency-dependent, monochromatic conductivity $\sigma(\omega)$ is given by

$$\sigma(\omega) = \int_0^\infty f(\tau)e^{i\omega\tau}\,d\tau. \tag{35.3}$$

For the monochromatic components of $\mathbf{E}$ and $\mathbf{H} = \mathbf{B}/\mu_0$, the equations $(35.1)_{1,3}$ now become

$$\mathbf{curl\ H} = \sigma(\omega)\mathbf{E}, \qquad \mathbf{curl\ E} = i\omega\mu_0\mathbf{H}. \tag{35.4}$$

If $\sigma(\omega)$ is uniform throughout the conductor, $\operatorname{div}\sigma\mathbf{E} = \sigma\operatorname{div}\mathbf{E} = 0$ is a consequence of $(35.4)_1$. Similarly, $\operatorname{div}\mathbf{H} = 0$ follows from $(35.4)_2$. If we eliminate $\mathbf{H}$ between the two equations (35.4), we obtain

$$\mathbf{\Delta E} + k^2\mathbf{E} = 0, \qquad k^2 = i\omega\mu_0\sigma, \tag{35.5}$$

where we have used div $\mathbf{E} = 0$ in order to replace $\mathbf{curl}^2 \mathbf{E}$ by $-\mathbf{\Delta E}$. To (35.5) we must add the equation

$$\text{div}\, \mathbf{E} = 0, \qquad (35.6)$$

which, we recall, is a consequence of (35.4)$_1$, but not of (35.5). It is immediately verified that $\mathbf{H}$, too, satisfies (35.5)–(35.6).

We apply these equations to a conducting wire of circular cross section. Using cylindrical coordinates (r, θ, z), we seek a solution with $\mathbf{E} = (0, 0, E(r))$. Outside the wire, the electric field will be uniform and equal to the field $E(a)$ at the surface $r = a$ of the wire. Inside, $E(r)$ will be the solution of

$$\frac{1}{r}\frac{d}{dr}\left(r\frac{dE}{dr}\right) + k^2 E = 0, \qquad k = \frac{1+i}{\delta}, \qquad \delta = \sqrt{\frac{2}{\omega\mu_0\sigma}}, \qquad (35.7)$$

which is finite at $r = 0$. The form of the solution (known as a Bessel function) can be obtained in the following way.

Near the axis, we expand $E(r)$ as a power series $\sum a_n r^n$, which we substitute in (35.7). The result is

$$\sum_{n=1} n^2 a_n r^{n-2} + k^2 \sum_{n=0} a_n r^n = 0. \qquad (35.8)$$

It follows that $a_1 = 0$ and $a_m = -k^2 a_{m-2}/m^2$ for $m = 1, 2, \ldots$. The expansion therefore contains only even powers, and the electric field near the axis has the form

$$E = \text{const.} \times \left[1 - \tfrac{1}{2}i(r/\delta)^2 - \tfrac{1}{16}(r/\delta)^4\right] e^{-i\omega t}. \qquad (35.9)$$

The amplitude of E, and with it that of the current density, increases away from the axis of the wire as $1 + \tfrac{1}{8}(r/\delta)^4$. The constant in (35.9) is determined by the amplitude of the electric field at $r = a$ (say), or by the total current corresponding to the current density $\mathbf{j} = \sigma\mathbf{E}$. The approximate formula (35.9) is valid for $r \ll \delta$. For low frequencies, such that $a \ll \delta$, it holds up to the boundary.

Near the boundary $r = a$ of the wire, we neglect the curvature of the boundary and regard it as a plane boundary. Then (35.7)$_1$ becomes $E''(r) + k^2 E = 0$, and the solution that remains finite for $r < a$ is

$$E = \text{const.} \times e^{-(a-r)/\delta} e^{i(a-r)/\delta} e^{-i\omega t}. \qquad (35.10)$$

This approximation at the boundary is valid if the frequency is large enough so that $\delta \ll a$. The electric field, and with it the current density, is then confined to a thin layer of thickness δ near the surface of the wire. This is the *skin effect*. Since the effective cross section of the conducting wire is reduced, its resistance increases at high frequencies. The same is true of the Joule warming at a given total current. In the limit of very high ω or σ we have $\delta \to 0$: the current becomes a surface current and the electromagnetic field ($\mathbf{E}$ and $\mathbf{B}$) vanishes inside the conductor. In other words, the conductor behaves like a superconductor.

36. Magnetohydrodynamics

In the foregoing sections we have assumed the conductor to be stationary. If it is moving (with respect to an aether frame), Ohm's law (32.2) gives $\mathbf{j} - q\dot{\mathbf{x}} = \sigma(\mathbf{E} + \dot{\mathbf{x}} \times \mathbf{B})$ and we must add $\mathrm{div}(q\dot{\mathbf{x}} + \sigma\dot{\mathbf{x}} \times \mathbf{B})$ to the left hand side of equation (32.4). The charge density will then, in general, relax to a non-vanishing distribution. If we now restrict ourselves to pure conductors (cf. §15) and substitute the relations $\mathbf{H} = \mathbf{B}/\mu_0$, $\mathbf{D} = \epsilon_0 \mathbf{E}$, $\mathbf{j} = \mathcal{J} + q\dot{\mathbf{x}}$ and $\mathbf{E} + \dot{\mathbf{x}} \times \mathbf{B} = \mathcal{J}/\sigma$ into $\mathrm{curl}\,\mathbf{H} = \mathbf{j} + \mathbf{D}_t$, we obtain

$$\mathrm{curl}\,\mathbf{B}/\mu_0 = \mathcal{J} + \epsilon_0 \dot{\mathbf{x}}\,\mathrm{div}(\mathcal{J}/\sigma - \dot{\mathbf{x}} \times \mathbf{B}) + \epsilon_0(\mathcal{J}/\sigma - \dot{\mathbf{x}} \times \mathbf{B})_t. \quad (36.1)$$

There are five terms on the right hand side of this equation. We shall now introduce a series of assumptions under which the last four terms may be dropped. At any point in the conducting medium, let ℓ be the smallest length over which the various quantities typically vary; then the derivative of any quantity f may be estimated as f/ℓ (at most). Let us now assume that the following inequalities hold:

$$\epsilon_0\mu_0\dot{x}^2 \ll 1, \quad \epsilon_0\mu_0\omega\ell\dot{x} \ll 1, \quad \frac{\omega\epsilon_0}{\sigma} \ll 1, \quad \frac{1}{(\mu_0 c\ell\sigma)^2} \ll 1. \quad (36.2)$$

The first inequality states that the squared velocity (of motion with respect to the aether) is small compared to c^2. This assumption of non-relativistic motion lies at the basis of our thermodynamic theory of macroscopic bodies. The second inequality is a similar restriction on the product of $\omega\ell$ (which has the dimensions of a velocity) and $\dot{\mathbf{x}}$. The third is the one we have already assumed above (for a pure conductor we replace ϵ by ϵ_0); it ensures that charge relaxation occurs more rapidly than any other change. The last inequality concerns the product of the length ℓ and the conductivity σ. In a metallic conductor, it requires that ℓ should be larger than an angstrom. In a cloud of astronomical dimensions, say 10^{15} m, it would be satisfied even if the conductivity were that of a terrestrial piece of glass. The inequalities (36.2) may be thought of as characterizing non-relativistic, low frequency phenomena in a good conductor. The second and third tell us what 'low frequency' means; the third and fourth, what constitutes a 'good' conductor of electricity.

It is now easy to check that, whenever the inequalities (36.2) hold, the last four terms on the right hand side of (36.1) are negligible. This is not all, for the inequalities have further consequences (again, these are easily verified): convection current densities are negligible, both in Ohm's law $\mathbf{j} - q\dot{\mathbf{x}} = \sigma\mathcal{E}$ and in the jump condition $\mathbf{n} \times [\![\mathbf{H}]\!] = \mathbf{K}$ (where $\mathbf{K}$ would properly include the convective part $\sigma\dot{\mathbf{x}}$ associated with the surface charge density σ); the electric terms in the stress tensor are small compared with the magnetic terms; the electric force density $q\mathcal{E}$ is small compared with $\mathcal{J} \times \mathbf{B}$ (which, to the same approximation, may be replaced by $\mathbf{j} \times \mathbf{B}$); and the electric part, $\epsilon_0 E^2/2$, of the energy density is small compared with the magnetic part, $B^2/(2\mu_0)$.

Low frequency phenomena in a good Ohmic conductor which is also a viscous fluid are then described by the following equations:

$$\mathbf{curl}\,\mathbf{B}/\mu_0 = \mathbf{j}, \qquad \mathbf{j} = \sigma(\mathbf{E} + \dot{\mathbf{x}} \times \mathbf{B}),$$

$$\mathrm{div}\,\mathbf{B} = 0, \qquad \mathbf{curl}\,\mathbf{E} = -\mathbf{B}_t,$$

$$T = -[p + B^2/(2\mu_0) + \zeta\,\mathrm{div}\,\dot{\mathbf{x}}]I + \mathbf{B} \otimes \mathbf{B}/\mu_0 + 2\lambda d,$$

$$\varepsilon = \varphi - \vartheta\varphi_\vartheta + \tfrac{1}{2}\dot{x}^2 + B^2/(2\mu_0\rho),$$

$$\rho\ddot{\mathbf{x}} = -\,\mathbf{grad}\,p + \mathbf{j} \times \mathbf{B} + (2\lambda + \zeta)\,\mathbf{grad}\,\mathrm{div}\,\dot{\mathbf{x}} - \lambda\,\mathbf{curl}^2\,\dot{\mathbf{x}} + \rho\mathbf{b},$$

$$\rho\vartheta\dot{\eta} = j^2/\sigma + 2\lambda\,\mathrm{tr}(d \cdot d) + \zeta(\mathrm{tr}\,d)^2 + \rho h - \mathrm{div}\,\mathbf{q}, \qquad (36.3)$$

where λ and ζ are the coefficients of viscosity (§17). The system (36.3) constitutes the laws of magnetohydrodynamics. The name signifies the preponderance of the magnetic terms, but it would be a mistake to conclude that the electric field is unimportant, let alone negligible. The equations $\operatorname{div}\epsilon_0 \mathbf{E} = q$ and $\mathbf{n} \cdot [\![\epsilon_0 \mathbf{E}]\!] = \sigma$ still hold, of course; they are omitted from (36.3) only because they may be *separated* from the other equations (since the convective current densities $q\dot{\mathbf{x}}$ and $\sigma\dot{\mathbf{x}}$ can be dropped everywhere). But wherever there is a vacuum, as there must be around any finite system of conductors, Ohm's law $(36.3)_2$ is in fact *replaced* by $\operatorname{div}\mathbf{E} = 0$.

PROBLEM

Problem 36–1. Deduce from (36.3), for constant σ, the equation

$$\mathbf{B}_t = \frac{1}{\mu_0 \sigma}\Delta\mathbf{B} + \operatorname{curl}(\dot{\mathbf{x}} \times \mathbf{B}). \tag{36.4}$$

For a stationary conductor, we obtain

$$\mathbf{B}_t = \frac{1}{\mu_0 \sigma}\Delta\mathbf{B}. \tag{36.5}$$

This is, mathematically speaking, a diffusion equation. For an isolated body it leads to a decay, or relaxation, of the magnetic field on a time scale of $\mu_0\sigma\ell^2/(2\pi)^2$. For example, if the earth's magnetic field has its sources in electric currents that flow in its conducting, molten core, the magnetic field should decay after a few times 10^4 years (based on $\sigma \approx 10^7$ siemen/m for the molten iron in the earth's core, and $\ell \approx 10^3$ km.) Since magnetically aligned sediments have been found that are *millions* of years old, we infer that the earth's interior must have been stirred, at least occasionally, because only the last term of (36.4) can halt the decay of the magnetic field.

As a non-trivial example of (36.3), we consider the steady, rectilinear flow (in the x direction) of an incompressible, conducting, viscous fluid between two parallel planes (at $z = \pm a$) in the presence of an imposed magnetic field B_0 in the direction normal to the planes.† In an incompressible fluid $\mathbf{grad}\, p$ can always be taken to include the force density

† Hartmann (1937)

$\rho\,\mathbf{grad}\,U = \mathbf{grad}\,\rho U$ of gravity. We shall assume all quantities, except for the pressure, to depend on z only; a pressure gradient (or a gravitational force) in the flow direction x may be necessary for maintaining the viscous flow. The condition $\text{div}\,\dot{\mathbf{x}} = 0$ for incompressible flow is satisfied by any velocity field $\dot{x}(z)$ in the x direction. For steady, incompressible flow, and with the assumed uniformity in the x and y directions, equations (36.3) give

$$j_y = \frac{1}{\mu_0}\frac{\partial B_x}{\partial z} = \sigma(E_y - \dot{x}B_z), \qquad j_x = j_z = 0,$$

$$B_z = B_0 \quad \text{and} \quad E_y \quad \text{are constants,}$$

$$0 = -\frac{\partial p}{\partial x} + j_y B_z + \lambda\Delta\dot{x}, \qquad 0 = -\frac{\partial p}{\partial z} - j_y B_x, \tag{36.6}$$

where we have used (17.5) and $\mathbf{curl}^2 = \mathbf{grad}\,\text{div} - \mathbf{\Delta}$. From the first, second and last members of (36.6) we conclude that $p + B_x^2(z)/(2\mu_0)$ is independent of z. Hence $\partial p/\partial x$ is independent of z. By eliminating j_y we obtain

$$\lambda\frac{\partial^2\dot{x}}{\partial z^2} - \sigma B_0^2\dot{x} = -\frac{\partial p}{\partial x} - \sigma E_y B_z. \tag{36.7}$$

This is an ordinary differential equation with constant coefficients (and constant inhomogeneous term). We impose the boundary condition that $\dot{x}$ is to vanish at $z = \pm a$, where the viscous fluid is assumed to adhere to the walls.

PROBLEM

Problem 36–2. Show that the solution is

$$\dot{x} = \dot{x}_0\frac{\cosh(a/h) - \cosh(z/h)}{\cosh(a/h) - 1}, \tag{36.8}$$

where $h = (\lambda/\sigma)^{1/2}/B_0$ and $\dot{x}_0$, the velocity on the mid-plane $z = 0$, is proportional to the right hand side of (36.7).

The velocity profile (36.8) depends on B_0 through the *Hartmann number* a/h. When $a/h \to 0$ (or $B_0 \to 0$), the velocity profile becomes parabolic:

$$\dot{x} = \dot{x}_0(1 - z^2/a^2). \tag{36.9}$$

When $a/h \to \infty$ (or $B_0 \to \infty$), the profile becomes flattened, except for boundary layers of thickness $\approx h$ along the planes:

$$\dot{x} = \dot{x}_0(1 - e^{(a-|z|)/h}).$$ (36.10)

The foregoing solution provides an approximation to the flow in a rectangular channel if the width in the y direction is large compared with $2a$. If we integrate $(36.6)_1$ with respect to z, we obtain

$$J_y = 2a\sigma(E_y - B_0\dot{x}_m),$$ (36.11)

where $J_y = \int j_y \, dz$ is the total lateral current per unit length of the channel and $\dot{x}_m = (2a)^{-1} \int \dot{x} \, dz$ is the mean velocity. The current J_y will depend on the external return path between the lateral planes ($y = \pm$const.). If this path is open, $J_y = 0$. The electric field component $E_y = B_0\dot{x}_m$ can be measured, and the arrangement provides an electromagnetic flow meter. If the path is short-circuited, $E_y = 0$ and $J_y = -2a\sigma B_0\dot{x}_m$. In other cases, depending on the external resistance and the pressure gradient $-\partial p/\partial x$, the device may function as a electromagnetic pump, brake or generator.

Although we have written down the equations (36.3) for fluids, the magnetohydrodynamic approximation (36.2) can be equally applied to elastic bodies; all that is necessary is the deletion, in (36.3), of the viscous terms and the replacement of the pressure p by $-\rho\varphi_F F^T$, and of $\mathbf{grad}\, p$ by $-\mathbf{div}\, \rho\varphi_F F^T$.

If, in addition to the inequalities (36.2), the more stringent inequalities

$$\mu_0\sigma\ell\dot{x} \gg 1, \qquad \mu_0\sigma\ell^2\omega \gg 1$$ (36.12)

hold, the first term on the right hand side of (36.4) becomes negligible, and the material behaves like a perfect conductor. Ohm's law $\mathcal{E} = \mathbf{j}/\sigma$ then reduces to $\mathcal{E} = 0$, and the Joule warming term j^2/σ in $(36.3)_8$ becomes negligible.

PROBLEMS

In the following problems the fluid is assumed to be non-viscous.

Problem 36–3. Prove the identity

$$(\mathrm{grad}\, \dot{\mathbf{x}})\dot{\mathbf{x}} = (\mathbf{curl}\, \dot{\mathbf{x}}) \times \dot{\mathbf{x}} + \mathbf{grad}\, \tfrac{1}{2}\dot{x}^2.$$ (36.13)

Hence show that $(36.3)_7$ can be written in the form

$$\rho[\dot{\mathbf{x}}_t + (\mathbf{curl}\ \dot{\mathbf{x}}) \times \dot{\mathbf{x}} + \mathbf{grad}\ \tfrac{1}{2}\dot{x}^2] = -\mathbf{grad}\ p$$

$$+(\mathbf{curl}\ \mathbf{B}) \times \mathbf{B}/\mu_0 + \rho\mathbf{b}. \tag{36.14}$$

Problem 36–4. Let $\mathbf{B}$ be any solenoidal field. Show that, if $\mathbf{b} = -\mathbf{grad}\ U$, solutions of the magnetohydrodynamic equations (36.3) for a perfectly conducting, incompressible fluid are obtained by choosing

$$\dot{\mathbf{x}} = \frac{\pm\mathbf{B}}{\sqrt{\mu_0\rho}} \qquad \text{and} \qquad p + \tfrac{1}{2}\rho\dot{x}^2 + \rho U = \text{const.}$$

Problem 36–5. Let $\mathbf{B}$ be a solenoidal field which becomes uniform, $\mathbf{B} \to \mathbf{B}_\infty$, at infinity. For a perfectly conducting, incompressible fluid with a conservative body force, the simple solutions of Problem 36–4 tend to a constant flow, $\pm\mathbf{u} = \pm\mathbf{B}_\infty/\sqrt{\mu_0\rho}$, at infinity. By applying a Galilei transformation $\mathbf{x}' = \mathbf{x} - \mathbf{u}t$, show that the solutions in the (x') frame become

$$\mathbf{B}' = \mathbf{B}(\mathbf{x}' \pm \mathbf{u}t),$$

$$\dot{\mathbf{x}}' = \pm\frac{\mathbf{B}(\mathbf{x}' \pm \mathbf{u}t)}{\sqrt{\mu_0\rho}} \mp \mathbf{u}.$$

These are finite-amplitude waves (called *Alfvén waves*) travelling at the constant velocity $\pm\mathbf{u} = \pm\mathbf{B}_\infty/\sqrt{\mu_0\rho}$.

37. The Faraday disk

We conclude this chapter with a detailed example of a moving, rigid conductor: the *Faraday disk*, also called the *unipolar inductor*, or the *homopolar generator*. It consists of a non-magnetic hollow cylinder, or disk, which rotates with constant angular velocity Ω in a uniform magnetic field $\mathbf{B}$, parallel to the axis. Sliding contacts are attached to the inner and outer surfaces of the disk, and they are connected through an external resistance (Figure 14). The disk, of conductivity σ, has inner radius a, outer radius b and thickness L. We assume the inner and outer surfaces, and the sliding contacts, to be highly conducting. Then $r = a$ and $r = b$ are equipotentials, and the electric field in the disk is radial. Since $\dot{\mathbf{x}} \times \mathbf{B}$ is also radial, so are $\mathcal{E} = \mathbf{E} + \dot{\mathbf{x}} \times \mathbf{B}$ and $\mathbf{j} = \sigma\mathcal{E}$. According to (36.3), we have, in this steady case,

$$\text{div } \mathbf{j} = 0,$$

$$\mathbf{j} = \sigma(\mathbf{E} + \dot{\mathbf{x}} \times \mathbf{B}),$$

$$\mathbf{curl}\ \mathbf{E} = 0. \tag{37.1}$$

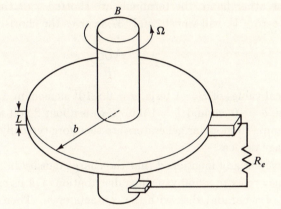

FIGURE 14

These equations are the same as (33.1), with $\mathbf{j} = \sigma\mathbf{E}$ replaced by $\mathbf{j} = \sigma\boldsymbol{\mathcal{E}}$.
From them we obtain, respectively,

$$j = \frac{i}{2\pi r L}, \qquad j = \sigma(E + \Omega r B),$$

$$\int_a^b E\, dr - V_e = 0, \tag{37.2}$$

where V_e is the voltage across the external resistance. Hence

$$V_e = \int_a^b \left(\frac{i}{2\pi\sigma Lr} - \Omega r B \right) dr = \frac{i}{2\pi\sigma L} \ln \frac{b}{a} - \tfrac{1}{2}\Omega B(b^2 - a^2)$$

$$= Ri - \frac{\Omega}{2\pi} B\pi(b^2 - a^2), \tag{37.3}$$

where R is the resistance (33.11) of the disk.

We consider two extreme cases. If the circuit is open – or the external resistance infinite – the current will vanish, and (37.3) gives the open circuit voltage as

$$V_{OC} = -\frac{\Omega}{2\pi} B\pi(b^2 - a^2). \tag{37.4}$$

If, on the other hand, the terminals are shorted – or the external resistance zero – V_e will vanish, and (37.3) gives the short-circuit current as

$$i_{SC} = \frac{V_{OC}}{R}. \tag{37.5}$$

For typical values of $B = 1$ tesla, $\sigma = 6 \times 10^7$ siemen/m, $\Omega = 3600$ rpm, $a = 1$ cm, $b = 10$ cm and $L = 1$ mm, V_{OC} is about 2 volt and i_{SC} about 3×10^5 amp. Homopolar generators are therefore typically high-current, low-voltage devices.

We have already mentioned that the charge density in a moving conductor may relax to a non-vanishing distribution. Let us examine this in the case of a Faraday disk with the contacts open. Then $\boldsymbol{\mathcal{J}} = \sigma\boldsymbol{\mathcal{E}} = 0$, i.e. $\mathbf{E} = -\dot{\mathbf{x}} \times \mathbf{B}$, and the charge density is $q = \operatorname{div} \epsilon_0 \mathbf{E} = -2\epsilon_0 \boldsymbol{\Omega} \cdot \mathbf{B}$, a negative constant. The total volume charge is $-2\epsilon_0 \Omega B\pi(b^2 - a^2)L$. The electric field must vanish for $r < a$; otherwise it would become singular at $r = 0$. Hence the surface charge density at $r = a$ is $\sigma_a = \epsilon_0 E_a = -\epsilon_0 \Omega a B$. If the disk was neutral to begin with, the surface charge density at $r = b$ must be such that the total surface charge $\int \sigma \, dS$ be equal to $\int -q \, dV$. This gives $\sigma_b = \epsilon_0 \Omega b B$. The electric field then vanishes outside. With the surface charge densities there will be associated surface current densities $\sigma_a \Omega a = -(\Omega a/c)^2 B/\mu_0$ and $\sigma_b \Omega b = (\Omega b/c)^2 B/\mu_0$. The jumps $\mathbf{n} \times [\![\mathbf{H}]\!] = \mathbf{n} \times [\![\mathbf{B}]\!]/\mu_0$ at the boundaries will therefore vanish to non-relativistic accuracy. Similarly, the convection current $q\dot{\mathbf{x}}$ in

curl B $= q\dot{\mathbf{x}}$ is relativistically small. This justifies our (hitherto tacit) assumption that **B** in the disk is the same as outside.

If the circuit is closed, the volume force $\mathbf{j} \times \mathbf{B}$ has a torque, relative to a point on the axis, of magnitude $\frac{1}{2}iB(b^2 - a^2)$. An external torque, of equal magnitude but opposite sign, must then be applied in order to maintain the angular velocity Ω of the generator. It is also possible to *impose* on the circuit an externally maintained current i (in the opposite direction); the magnetic torque will then turn the disk, and the homopolar generator will become a *motor*.

PROBLEM

Problem 37–1. A conducting sphere of radius a is placed in a uniform magnetic field **B** and made to rotate with angular velocity Ω around an axis that is parallel to the field. Find the electric potential inside and outside the sphere.

CHAPTER X

Radiation

38. The wave equations

The second pair of Maxwell's equations can always be satisfied by introducing the potentials $\mathbf{A}$ and V:

$$\mathbf{B} = \operatorname{curl} \mathbf{A},$$

$$\mathbf{E} = -\mathbf{A}_t - \operatorname{grad} V. \tag{38.1}$$

In an aether frame, the first pair of Maxwell's equations is (cf. (8.8))

$$\operatorname{div} \epsilon_0 \mathbf{E} = q,$$

$$\operatorname{curl} \mathbf{B}/\mu_0 - \epsilon_0 \mathbf{E}_t = \mathbf{j}, \tag{38.2}$$

where q and $\mathbf{j}$ are the *total* charge and current densities, including those of polarization and magnetization. For linear dielectrics and magnets, we may replace ϵ_0 and μ_0 by ϵ and μ, respectively, and then q and $\mathbf{j}$ are the free charge and current densities; we shall then also have to replace, in the following equations, the squared speed of light in vacuum $c^2 = (\epsilon_0 \mu_0)^{-1}$ by $(\epsilon \mu)^{-1}$. If we substitute (38.1) in (38.2) we obtain

$$\Delta V - \frac{1}{c^2} V_{tt} + \frac{\partial}{\partial t}\left(\operatorname{div} \mathbf{A} + \frac{1}{c^2} V_t\right) = -\frac{q}{\epsilon_0},$$

$$\Delta \mathbf{A} - \frac{1}{c^2} \mathbf{A}_{tt} - \operatorname{grad}\left(\operatorname{div} \mathbf{A} + \frac{1}{c^2} V_t\right) = -\mu_0 \mathbf{j}. \tag{38.3}$$

The potentials are still subject to a gauge transformation, and this can be chosen so that they satisfy the *Lorenz gauge condition*

$$\text{div } \mathbf{A} + \frac{1}{c^2} V_t = 0. \tag{38.4}$$

Equations (38.3) then become

$$\Delta V - \frac{1}{c^2} V_{tt} = -\frac{q}{\epsilon_0},$$

$$\Delta \mathbf{A} - \frac{1}{c^2} \mathbf{A}_{tt} = -\mu_0 \mathbf{j}. \tag{38.5}$$

These are inhomogeneous wave equations for V and $\mathbf{A}$. If the time derivatives are discarded, the Lorenz gauge (38.4) reduces to the Coulomb gauge (27.3); (38.5)$_1$ reduces to Poisson's equation (19.3); and (38.5)$_2$ reduces to (27.4).

The strangest claims have been made regarding the solutions of the inhomogeneous wave equations (38.5), usually because of careless mathematics. We shall therefore treat these equations in a mathematically responsible manner. This requires a little patience.

Each of the four components of (38.5) is of the form

$$\Delta u - \frac{1}{c^2} u_{tt} = -g, \tag{38.6}$$

which is a linear partial differential equation for u. Solutions of this *inhomogeneous* equation can be obtained from any *particular* one by adding solutions of the *homogeneous* equation with $g = 0$. This is, of course, the analogue of the statement that solutions of Poisson's equation $\Delta V = -q/\epsilon_0$ can be obtained from any particular one by adding harmonic functions. But in the present case there are initial conditions to consider.

We begin with a discussion of the homogeneous wave equation,

$$\Delta u - \frac{1}{c^2} u_{tt} = 0. \tag{38.7}$$

Solutions of this equation that have the form $u(\mathbf{x}, t) = f(\mathbf{x}) g(h(\mathbf{x}) \pm ct)$ are called *progressive waves*, or *travelling waves*. The argument $h(\mathbf{x}) \pm ct$ of g is called the *phase* of the wave; the surfaces $h(\mathbf{x}) = $ const. are called the *wave fronts*. Solutions of the form $u(\mathbf{x}, t) = f(\mathbf{x}) g(t)$ are called *standing waves.* The most important example of progressive waves is given by the spatially decreasing spherical waves: if we look for solutions of (38.7) that depend only on $r = |\mathbf{x}|$ and t, that is $u = u(r, t)$, we find $\Delta u = u_{rr} + (2/r)u_r = (ru)_{rr}/r = (1/c^2)u_{tt}$, or $(ru)_{rr} - (1/c^2)(ru)_{tt} = 0$. It is easy to show that the general solution of the latter equation is

$$u(r, t) = \frac{f_1(r - ct)}{r} + \frac{f_2(r + ct)}{r}, \qquad (38.8)$$

where f_1 and f_2 are arbitrary. The first term is an outgoing, the second an incoming, spherical wave. Of course we may choose to define r as the distance from any point P; then (38.8) is still the general solution, and it describes waves which emanate from, or converge to, P. These spherical waves play a role which is similar to that of the fundamental harmonic function $1/r$ in electrostatics.

Let $r = |\mathbf{y} - \mathbf{x}|$, the distance from $\mathbf{x}$ to $\mathbf{y}$. Then, for any ψ and f,

$$u(\mathbf{x}, t) = \int \psi(\mathbf{y}) \frac{f(r - ct)}{r} \, d^3 y \qquad (38.9)$$

is a superposition of spherical waves issuing from $\mathbf{x}$, and is therefore a solution of the wave equation (38.7). We choose $f(\lambda)$ to be a non-negative function that vanishes outside the interval $-\epsilon < \lambda < \epsilon$, and for which

$$\int_{-\infty}^{\infty} f(\lambda) \, d\lambda = 1.$$

The contributions to the integral in (38.9) are then only from a spherical shell of thickness 2ϵ around $r = ct$. Letting ϵ tend to zero and passing to the limit, so that f becomes a delta function, we obtain

$$u = ct \int_{|\mathbf{y}-\mathbf{x}|=ct} \psi(\mathbf{y}) d\omega, \qquad (38.10)$$

where $d\omega$ is an element of solid angle. Any point on the sphere $|\mathbf{y} - \mathbf{x}| = ct$ can be written as $\mathbf{y} = \mathbf{x} + \mathbf{n}ct$, where $\mathbf{n}$ is a unit radial vector. We denote

$$M(t)\psi = \frac{1}{4\pi} \int \psi(\mathbf{x} + \mathbf{n}ct) \, d\omega. \qquad (38.11)$$

This is a mean of ψ over the sphere of radius ct around $\mathbf{x}$. It is therefore a function of $\mathbf{x}$ and t, and its value at $t = 0$ is $\psi(\mathbf{x})$. We now compute its time-derivative:

$$(M(t)\psi)_t = \frac{c}{4\pi} \int \mathbf{n} \cdot \mathbf{grad}\, \psi \, d\omega$$

$$= \frac{1}{4\pi c t^2} \int_{|\mathbf{y}-\mathbf{x}|=ct} \mathbf{n} \cdot \mathbf{grad}\, \psi(\mathbf{y}) \, dS$$

$$= \frac{1}{4\pi c t^2} \int_{|\mathbf{y}-\mathbf{x}|\leq ct} \Delta \psi(\mathbf{y}) \, d^3 y, \qquad (38.12)$$

where we have used $dS = r^2 \, d\omega = c^2 t^2 \, d\omega$, as well as Gauss's theorem. Since the volume in the last integral goes as $c^3 t^3$, $(M(t)\psi)_t$ vanishes at $t = 0$.

According to (38.10) and (38.11), the function

$$u(\mathbf{x}, t) = t M(t)\psi, \qquad (38.13)$$

for any ψ, is a solution of the homogeneous wave equation (38.7). Since $u_t = t(M(t)\psi)_t + M(t)\psi$, u satisfies the initial conditions

$$u(\mathbf{x}, 0) = 0, \qquad u_t(\mathbf{x}, 0) = \psi(\mathbf{x}). \qquad (38.14)$$

With this u, let $v = u_t$. Then v, too, satisfies the wave equation and has the initial values $v(\mathbf{x}, 0) = u_t(\mathbf{x}, 0) = \psi(\mathbf{x})$ and $v_t(\mathbf{x}, 0) = u_{tt}(\mathbf{x}, 0) = c^2 \Delta u(\mathbf{x}, 0) = 0$. Since ψ is arbitrary, this proves that, for any φ, the function

$$u(\mathbf{x}, t) = (t M(t)\varphi)_t \qquad (38.15)$$

is a solution of (38.7) with the initial values

$$u(\mathbf{x}, 0) = \varphi(\mathbf{x}), \qquad u_t(\mathbf{x}, 0) = 0. \qquad (38.16)$$

By superposing these solutions, we conclude that

$$u(\mathbf{x}, t) = (t M(t)\varphi)_t + t M(t)\psi \qquad (38.17)$$

is a solution with initial values

$$u(\mathbf{x}, 0) = \varphi(\mathbf{x}), \qquad u_t(\mathbf{x}, 0) = \psi(\mathbf{x}). \qquad (38.18)$$

We now turn to the inhomogeneous wave equation and, using an idea known as *Duhamel's principle*, we seek a particular solution with zero initial conditions. Having found it, we can always make it conform to the initial conditions (38.18) by adding (38.17).

Consider the following initial value problem at $t = \tau$:

$$\Delta v - \frac{1}{c^2} v_{tt} = 0, \qquad v(\mathbf{x}, \tau) = 0, \qquad v_t(\mathbf{x}, \tau) = g(\mathbf{x}, \tau). \qquad (38.19)$$

According to (38.13) and (38.14), the solution of (38.19), which we denote by $v(\mathbf{x}, t; \tau)$, is

$$v(\mathbf{x}, t; \tau) = (t - \tau) M(t - \tau) g. \qquad (38.20)$$

We now prove that

$$u(\mathbf{x}, t) = c^2 \int_0^t v(\mathbf{x}, t; \tau) \, d\tau \qquad (38.21)$$

is a solution of the inhomogeneous equation (38.6) with zero initial conditions. For

$$u_t = c^2 v(\mathbf{x}, t; t) + c^2 \int_0^t v_t(\mathbf{x}, t; \tau) \, d\tau$$

$$= c^2 \int_0^t v_t(\mathbf{x}, t; \tau) d\tau, \qquad (38.22)$$

since $v(\mathbf{x}, t; t) = 0$. Clearly, $u(\mathbf{x}, 0) = u_t(\mathbf{x}, 0) = 0$. Furthermore,

$$u_{tt} = c^2 v_t(\mathbf{x}, t; t) + c^2 \int_0^t v_t t(\mathbf{x}, t; \tau) \, d\tau$$

$$= c^2 g(\mathbf{x}, t) + c^2 \int_0^t v_t t d\tau, \qquad (38.23)$$

and

$$\Delta u = c^2 \int_0^t \Delta v \, d\tau. \qquad (38.24)$$

Hence $\Delta u - (1/c^2) u_{tt} = -g$.

In order to obtain a more illuminating form of this solution, we substitute (38.20) in (38.21) and recall the definition (38.11) of M:

$$u(\mathbf{x}, t) = c^2 \int_0^t \frac{d\tau (t - \tau)}{4\pi} \int g(\mathbf{x} + \mathbf{n} c(t - \tau), \tau) \, d\omega.$$

With a change of variable from τ to $r = c(t - \tau)$, we find

$$u(\mathbf{x}, t) = \frac{1}{4\pi} \int_0^{ct} r\, dr \int g(\mathbf{x} + \mathbf{n}r, t - r/c)\, d\omega$$

$$= \frac{1}{4\pi} \int_{r \leq ct} \frac{g(\mathbf{x}', t - r/c)\, d^3 x'}{r}, \qquad (38.25)$$

where r is the distance from $\mathbf{x}$ to the integration point $\mathbf{x}'$, and the integral extends over the sphere of radius ct around $\mathbf{x}$.

In the form given by the last line of (38.25), the solution is easy to visualize. Unlike the solution of Poisson's equation, each volume element $d^3 x'$ makes its contribution $g(\mathbf{x}', t - r/c)\, d^3 x'/r$ at the *retarded* time $t - r/c$, rather than at t; and the amount of retardation is precisely that which a signal, travelling at the velocity of light c, would need to traverse the distance r from $\mathbf{x}'$ to $\mathbf{x}$. In addition, only sources that are within a distance ct from $\mathbf{x}$ contibute to the integral. Those that lie beyond the sphere $r = ct$, which may be called the *horizon* at time t, make no contribution – except at a later time, when they rise, so to speak, above the horizon.

For a finite system (such that g vanishes outside a finite region of space), the solution (38.25), which is zero initially, will continue to vanish until such time as is required for the horizon to reach the system. If d is the distance from $\mathbf{x}$ to the nearest part of the system, this time is $t_1 = d/c$. The *whole* system will affect the solution after the time $t_2 = D/c$, where D is the distance to the furthest part of the system. After t_2 the whole system lies within the horizon, and the limitation $r \leq ct$ on the volume integral of (38.25) becomes irrelevant.

As an example, we consider a conducting sphere of radius a, on which the uniform charge density $\sigma(t)$ is made to vary so that $\sigma = 0$ for $t < 0$ and $\sigma(t) = \sigma_0 \sin \omega t$ for $t > 0$. At a distance R $(R > a)$ from the centre of the sphere, the electric potential (assumed to vanish up to $t = 0$) will vanish for $ct < R - a$. Later, it will be given by

$$V = \frac{1}{4\pi\epsilon_0} \int_{r \leq ct} \frac{\sigma(t - r/c)\, dS}{r}. \qquad (38.26)$$

The contribution to the integral from a ring (Figure 15) between r and $r + dr$ will be $\sigma_0 \sin \omega(t - r/c) 2\pi a \sin \theta a\, d\theta/r$, where $\theta(r)$ is given by $r^2 = a^2 + R^2 - 2aR\cos\theta$. Noting that $r\, dr = aR\sin\theta\, d\theta$, we obtain

$$V = \frac{a\sigma_0}{2\epsilon_0 R} \int \sin\omega(t - r/c)\, dr. \tag{38.27}$$

In the last integral, the lower limit is $R - a$, and the upper limit is the smaller of ct and $R + a$. An easy calculation gives

$$V = \frac{ac\sigma_0}{2\epsilon_0 \omega R}\left[1 - \cos\omega\left(t - \frac{R+a}{c}\right)\right], \quad R - a < ct < R + a,$$

$$V = \frac{ac\sigma_0}{\epsilon_0 \omega R} \sin\frac{\omega a}{c} \sin\omega\left(t - \frac{R}{c}\right), \quad R + a < ct. \tag{38.28}$$

We note that both expressions are of the form $f(t - R/c)/R$. Each one is therefore an outgoing spherical wave.

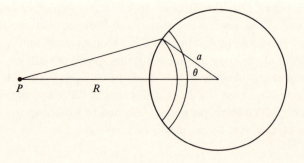

FIGURE 15

PROBLEM

Problem 38–1. Calculate the electric potential and field at a distance R $(R > a)$ from the centre of a sphere of radius a, which is suddenly given a (uniformly distributed) surface charge Q.

For a finite system, then, the solutions of equations (38.5) for the potentials are

$$\mathbf{A}(\mathbf{x}, t) = \frac{\mu_0}{4\pi} \int \frac{\mathbf{j}(\mathbf{x}', t - r/c)\, d^3 x'}{r},$$

$$V(\mathbf{x}, t) = \frac{1}{4\pi\epsilon_0} \int \frac{q(\mathbf{x}', t - r/c)\, d^3 x'}{r}, \qquad (38.29)$$

but these apply only after the time t_2, and for potentials that satisfy zero initial conditions.

Often, however, the formulae (38.29) are written down without mention of any conditions, and the claim is made that these *retarded potentials* are the 'appropriate' solutions of (38.5). Now, it is true that they are solutions, even for $t \leq t_2$. But they correspond to definite non-zero initial values of the potentials, which are determined by values of the sources at $(\mathbf{x}', -r/c)$, that is, in the past. It may be awkward, in practice, to adjust the initial conditions in accordance with this knowledge of the past distribution of the sources, but it is not impossible.

Along with these retarded potentials, there are the *advanced potentials*,

$$\mathbf{A}(\mathbf{x}, t) = \frac{\mu_0}{4\pi} \int \frac{\mathbf{j}(\mathbf{x}', t + r/c)\, d^3 x'}{r},$$

$$V(\mathbf{x}, t) = \frac{1}{4\pi\epsilon_0} \int \frac{q(\mathbf{x}', t + r/c)\, d^3 x'}{r}, \qquad (38.30)$$

which are also solutions, but they depend on the sources at $(\mathbf{x}', t + r/c)$, that is, in the *future*. They are often rejected as 'unphysical', because 'the effect cannot precede the cause', but the physicist should regard this piece of common wisdom with some criticism. Does he 'know' that the cause always precedes the effect? Is this a theorem that follows from the accepted principles of physics? If not, should it be added as a further principle? If so, how should it be formulated? In $\operatorname{div} \mathbf{D} = q$, is q the cause and $\mathbf{D}$ the effect, or is it the other way round? If the advanced solutions (38.30) of Maxwell's equations are unphysical, are not Maxwell's equations themselves unphysical?

The fact is that the whole issue of 'causality' is irrelevant. The advanced potentials are indeed solutions of Maxwell's equations, but they correspond to initial conditions that involve the sources at $(\mathbf{x}', r/c)$ – in the future. In order to arrange for such initial conditions we must therefore know the distribution of the sources in the future. In the absence of such knowledge, the advanced potentials are merely useless, not wrong.

PROBLEM

Problem 38–2. A uniform surface current $\mathbf{K}(t)$ is made to flow along an infinite plane. There is no electromagnetic field prior to $t = 0$. Find the electromagnetic field on either side of the plane for $t \geq 0$.

39. Radiation from finite systems

Instead of working directly with the formulae (38.29) for the potentials $\mathbf{A}$ and V, it often proves convenient to note that the Lorenz gauge condition,

$$\operatorname{div} \mathbf{A}/\mu_0 + \epsilon_0 V_t = 0, \tag{39.1}$$

has the same form as the law of charge conservation. It can therefore be satisfied by introducing two vectors $\mathbf{Z}^e$ and $\mathbf{Z}^m$, similar to the charge and current potentials $\mathbf{D}$ and $\mathbf{H}$:

$$\mathbf{A} = \mu_0(\mathbf{Z}_t^e - \operatorname{curl} \mathbf{Z}^m), \qquad V = -\epsilon_0^{-1} \operatorname{div} \mathbf{Z}^e. \tag{39.2}$$

The vectors $\mathbf{Z}^e$ and $\mathbf{Z}^m$ are called *Hertz vectors*. Since they are potentials for the potentials $\mathbf{A}$ and V, they are also called *superpotentials*. In terms of these, the fields (38.1) become

$$\mathbf{B} = \mu_0 \operatorname{curl} \mathbf{Z}_t^e - \mu_0 \operatorname{curl}^2 \mathbf{Z}^m,$$

$$\mathbf{E} = -\mu_0 \mathbf{Z}_{tt}^e + \epsilon_0^{-1} \operatorname{grad} \operatorname{div} \mathbf{Z}^e + \mu_0 \operatorname{curl} \mathbf{Z}_t^m, \tag{39.3}$$

and substitution in the first pair (38.2) of Maxwell's equations gives

$$\operatorname{div}\left(\mathbf{\Delta Z}^e - \frac{1}{c^2}\mathbf{Z}_{tt}^e\right) = q,$$

$$-\left(\mathbf{\Delta Z}^e - \frac{1}{c^2}\mathbf{Z}_{tt}^e\right)_t + \operatorname{curl}\left(\mathbf{\Delta Z}^m - \frac{1}{c^2}\mathbf{Z}_{tt}^m\right) = \mathbf{j}. \qquad (39.4)$$

Consider now the wave equations

$$\mathbf{\Delta Z}^e - \frac{1}{c^2}\mathbf{Z}_{tt}^e = -\mathbf{p},$$

$$\mathbf{\Delta Z}^m - \frac{1}{c^2}\mathbf{Z}_{tt}^m = \mathbf{m}, \qquad (39.5)$$

where $\mathbf{p}$ and $\mathbf{m}$, the *Hertz source vectors*, are given vector fields. The solutions of (39.5), at time t at a point P, which correspond to zero initial conditions, are

$$\mathbf{Z}^e = \frac{1}{4\pi}\int_{r \le ct} \frac{\mathbf{p}(\mathbf{x}, t - r/c)\, d^3 x}{r},$$

$$\mathbf{Z}^m = -\frac{1}{4\pi}\int_{r \le ct} \frac{\mathbf{m}(\mathbf{x}, t - r/c)\, d^3 x}{r}, \qquad (39.6)$$

where r is the magnitude of the vector $\mathbf{r}$ from the integration point $\mathbf{x}$ to P. From (39.4) and (39.5) it is clear that, for any pair of Hertz source vectors $\mathbf{p}$ and $\mathbf{m}$, the potentials (39.2) and the fields (39.3) that result from (39.6) are the solutions corresponding to the sources

$$q = -\operatorname{div}\mathbf{p}, \qquad \mathbf{j} = \mathbf{p}_t + \operatorname{curl}\mathbf{m}. \qquad (39.7)$$

Of course we know that, conversely, any charge-current distribution can be represented in this form (cf. (9.2)). In fact, the Hertz source vector $\mathbf{p}$ is often referred to as a *Hertz dipole*, and the source vector $\mathbf{m}$, as a *Hertz magnetic moment*; more precisely, $\mathbf{p}$ and $\mathbf{m}$ are *densities* (cf. (9.6)–(9.7)).

A simple example is provided by the Hertz source vectors $\mathbf{m} = 0$ and

$$\mathbf{p}(\mathbf{x}, t) = \mathbf{d}(t)\delta(\mathbf{x}), \qquad (39.8)$$

where $\delta(\mathbf{x})$ is the three-dimensional delta function. Since $\mathbf{p}$ is the dipole moment density and $\delta(\mathbf{x})$ is a unit density concentrated at the origin, (39.8) represents a time-dependent dipole $\mathbf{d}(t)$ at the origin. At any point with position vector $\mathbf{r}$ that, at time t, is within the horizon $r \le ct$, (39.6) give the solutions $\mathbf{Z}^m = 0$ and

$$\mathbf{Z}^e(\mathbf{r}, t) = \frac{1}{4\pi r}\mathbf{d}(t - r/c). \tag{39.9}$$

PROBLEM

Problem 39–1. Let $\mathbf{n} = \mathbf{r}/r$. Prove the following identities for $\mathbf{d}(t - r/c)$:

$$\operatorname{div} \mathbf{d} = -\frac{1}{c}\mathbf{n} \cdot \mathbf{d}_t,$$

$$\operatorname{div} \frac{1}{r}\mathbf{d} = -\frac{1}{r^2}\mathbf{n} \cdot \mathbf{d} - \frac{1}{cr}\mathbf{n} \cdot \mathbf{d}_t,$$

$$\operatorname{curl} \mathbf{d} = -\frac{1}{c}\mathbf{n} \times \mathbf{d}_t,$$

$$\operatorname{curl} \frac{1}{r}\mathbf{d} = -\frac{1}{r^2}\mathbf{n} \times \mathbf{d} - \frac{1}{cr}\mathbf{n} \times \mathbf{d}_t,$$

$$\operatorname{grad} \mathbf{r} \cdot \mathbf{d} = \mathbf{d} - \frac{r}{c}(\mathbf{n} \cdot \mathbf{d}_t)\mathbf{n}, \tag{39.10}$$

where $\mathbf{d}$ and its derivatives are, of course, evaluated at the retarded time $t - r/c$. Hence prove that the electromagnetic potentials (39.2) and fields (39.3) that follow from (39.9) are:

$$\mathbf{A} = \frac{\mu_0}{4\pi r}\mathbf{d}_t,$$

$$V = \frac{1}{4\pi\epsilon_0}\left(\frac{1}{r^2}\mathbf{n} \cdot \mathbf{d} + \frac{1}{cr}\mathbf{n} \cdot \mathbf{d}_t\right),$$

$$\mathbf{B} = -\frac{\mu_0}{4\pi}\left(\frac{1}{r^2}\mathbf{n} \times \mathbf{d}_t + \frac{1}{cr}\mathbf{n} \times \mathbf{d}_{tt}\right),$$

$$\mathbf{E} = \frac{1}{4\pi\epsilon_0}\{\frac{1}{r^3}[3(\mathbf{n} \cdot \mathbf{d})\mathbf{n} - \mathbf{d}] + \frac{1}{cr^2}[3(\mathbf{n} \cdot \mathbf{d}_t)\mathbf{n} - \mathbf{d}_t]\}$$

$$+ \frac{\mu_0}{4\pi r}[(\mathbf{n} \cdot \mathbf{d}_{tt})\mathbf{n} - \mathbf{d}_{tt}]. \tag{39.11}$$

Near the dipole, the electric field is dominated by its first term, which is an electrostatic field corresponding to the instantaneous value of the dipole moment $\mathbf{d}(t)$ (since the retardation vanishes at $r = 0$). Far away from the dipole, the fields are

$$\mathbf{E} = \frac{\mu_0}{4\pi} \frac{\mathbf{n} \times [\mathbf{n} \times \mathbf{d}_{tt}(t - r/c)]}{r},$$

$$\mathbf{B} = \frac{1}{c}\mathbf{n} \times \mathbf{E}. \tag{39.12}$$

In any given direction $\mathbf{n}$, the electric field in the far zone is an outgoing spherical wave. It is *transverse*, its direction being normal to $\mathbf{n}$, and its amplitude is angle-dependent. The magnetic field, too, is a transverse outgoing spherical wave; it is normal to both $\mathbf{n}$ and $\mathbf{E}$. The far zone is therefore also called the *wave zone*, or the *radiation zone*.

With any electromagnetic field there is associated an energy flux $\mathbf{E} \times \mathbf{H}$ (§12). In the far zone, where the fields are given by (39.12), this becomes

$$\mathbf{E} \times \mathbf{H} = \frac{\mathbf{E} \times \mathbf{B}}{\mu_0} = \frac{E^2}{\mu_0 c}\mathbf{n} = \frac{\mu_0}{4\pi} \frac{\sin^2\theta}{4\pi c r^2} d_{tt}^2 \mathbf{n}, \tag{39.13}$$

where θ is the angle between $\mathbf{d}$ and $\mathbf{n}$ (the polar angle measured from $\mathbf{d}$). If we therefore place (in the far zone) a target $dS = r^2\,d\Omega$ that is normal to $\mathbf{n}$, the amount of energy that enters it per unit time is

$$\mathbf{E} \times \mathbf{H} \cdot \mathbf{n}\,dS = \frac{1}{4\pi\epsilon_0} \frac{\sin^2\theta}{c^3} d_{tt}^2 \frac{d\Omega}{4\pi}. \tag{39.14}$$

This may be regarded as the power radiated by the dipole into the solid angle $d\Omega$. The angular distribution of the radiation is given by the factor $\sin^2\theta$: nothing in the fore and aft directions, and maximal power in the directions perpendicular to $\mathbf{d}$ (Figure 16).

By integrating (39.14) over all angles, we obtain Larmor's formula for the total power:

$$P = \int \mathbf{E} \times \mathbf{H} \cdot \mathbf{n}\,dS = \frac{1}{4\pi\epsilon_0} \frac{2}{3c^3} d_{tt}^2. \tag{39.15}$$

PROBLEM

Problem 39–2. Carry out the analogous calculation for a magnetic dipole, rep-

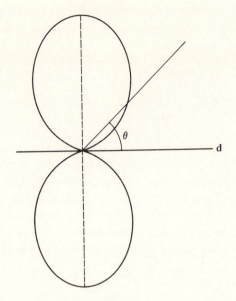

FIGURE 16

resented by the Hertz source vector

$$\mathbf{m}(\mathbf{x}, t) = \mathbf{d}(t)\delta(\mathbf{x}). \tag{39.16}$$

Prove the identities:

$$\mathbf{curl}\ \mathbf{r} \times \mathbf{d} = -2\mathbf{d} - \frac{r}{c}\mathbf{n} \times (\mathbf{n} \times \mathbf{d}_t),$$

$$\mathbf{curl}\ \frac{\mathbf{n} \times \mathbf{d}}{r^2} = \frac{1}{r^3}[\mathbf{d} - 3(\mathbf{n} \cdot \mathbf{d})\mathbf{n}] - \frac{1}{cr^2}\mathbf{n} \times (\mathbf{n} \times \mathbf{d}_t). \tag{39.17}$$

Show that the total power radiated is

$$P = \frac{\mu_0}{4\pi}\frac{2}{3c^3}d_{tt}^2. \tag{39.18}$$

We now apply our results for the Hertz electric dipole to the radiation from a linear antenna, represented by the segment $-L/2 \le z \le L/2$ on the z axis. Let

$$i = i_0 \cos \frac{\pi z}{L} \sin \omega t \qquad (39.19)$$

be an alternating current flowing in the antenna. If S is the cross section of the antenna, the law of charge conservation gives

$$\frac{1}{S}\frac{\partial i}{\partial z} = -q_t, \qquad (39.20)$$

from which we obtain the charge density:

$$q(z,t) = \frac{i_0\pi}{S\omega L} \sin \frac{\pi z}{L} \cos \omega t. \qquad (39.21)$$

At $t = 0$, q has a maximum at $z = L/2$ and a minimum at $z = -L/2$; at $t = \pi/(2\omega)$, q vanishes everywhere; and at $t = \pi/\omega$, the distribution of q is the same, except for sign, as at $t = 0$.

The antenna has the dipole moment

$$d(t) = \int zqS\,dz = \frac{2i_0 L}{\pi\omega} \cos \omega t \qquad (39.22)$$

in the z direction. If we regard it as a point dipole, the power it radiates is, according to (39.15),

$$P = \frac{1}{4\pi\epsilon_0}\frac{2}{3c^3}\left[\frac{2i_0\omega L}{\pi}\cos\omega(t - r/c)\right]^2. \qquad (39.23)$$

Averaged over a period $2\pi/\omega$, this becomes

$$\bar{P} = \frac{1}{4\pi\epsilon_0}\frac{4\omega^2 L^2 i_0^2}{3\pi^2 c^3}. \qquad (39.24)$$

The approximation involved in regarding the antenna as a point dipole is justified when $\omega L/(2\pi c)$, which is the length of the antenna divided by the wave-length $\lambda = 2\pi c/\omega$, is small.

40. Radiation from a moving point charge

In order to calculate the potential V for a moving point charge in accordance with $(38.29)_2$, we first take account of the retardation by integrating with respect to t' with the delta function $\delta(t' - t + |\mathbf{x} - \mathbf{x}'|/c)$:

$$V(\mathbf{x},t) = \frac{1}{4\pi\epsilon_0}\int\int \frac{q(\mathbf{x}',t')\delta(t' - t + |\mathbf{x} - \mathbf{x}'|/c)\,dt'\,d^3x'}{|\mathbf{x} - \mathbf{x}'|}. \qquad (40.1)$$

In this integral we now substitute the charge density

$$q(\mathbf{x}, t) = e\delta[\mathbf{x} - \mathbf{r}_0(t)] \tag{40.2}$$

that corresponds to a point charge with charge e which is moving along the trajectory $\mathbf{r}_0(t)$. The space integration gives

$$V(\mathbf{x}, t) = \frac{1}{4\pi\epsilon_0} \int \frac{e\delta[t' - t + |\mathbf{x} - \mathbf{r}_0(t')|/c]\, dt'}{|\mathbf{x} - \mathbf{r}_0(t')|}. \tag{40.3}$$

As before, we denote by

$$\mathbf{r}(\mathbf{x}, t) = \mathbf{x} - \mathbf{r}_0(t) \tag{40.4}$$

the vector from the position $\mathbf{r}_0(t)$ of the point charge to $\mathbf{x}$. Also, let

$$t' + \frac{r(\mathbf{x}, t')}{c} = \tau. \tag{40.5}$$

We regard this as (implicitly) determining t' as a function $t'(\mathbf{x}, \tau)$ of $\mathbf{x}$ and τ. By differentiation at fixed $\mathbf{x}$, we obtain

$$(1 - \mathbf{n} \cdot \mathbf{v}/c)\frac{\partial t'}{\partial \tau} = 1, \tag{40.6}$$

where

$$\mathbf{n} = \frac{\mathbf{r}(\mathbf{x}, t')}{r(\mathbf{x}, t')}, \qquad \mathbf{v} = \dot{\mathbf{r}}_0(t'). \tag{40.7}$$

We can now use (40.6) to change the variable of integration in (40.3) from t' to τ:

$$V(\mathbf{x}, t) = \frac{1}{4\pi\epsilon_0} \int \frac{e\delta(\tau - t)\, d\tau}{r(\mathbf{x}, t')(1 - \mathbf{n} \cdot \mathbf{v}/c)} = \frac{1}{4\pi\epsilon_0}\frac{e}{r - \mathbf{r} \cdot \mathbf{v}/c}, \tag{40.8}$$

where the denominator in the final result is to be taken at the value $t'(\mathbf{x}, t)$ of t' that satisfies (40.5) with $\tau = t$:

$$t' + \frac{r(\mathbf{x}, t')}{c} = t \qquad \text{or} \qquad r(\mathbf{x}, t') = c(t - t'). \tag{40.9}$$

A similar calculation of $\mathbf{A}$, starting with $(38.29)_1$ and

$$\mathbf{j}(\mathbf{x}, t) = q(\mathbf{x}, t)\mathbf{v}(t) = e\delta[\mathbf{x} - \mathbf{r}_0(t)]\dot{\mathbf{r}}_0(t), \tag{40.10}$$

leads to the result

$$\mathbf{A}(\mathbf{x}, t) = \frac{\mu_0}{4\pi}\frac{e\mathbf{v}}{r - \mathbf{r} \cdot \mathbf{v}/c}, \tag{40.11}$$

where, again, all quantities on the right hand side are evaluated at the 'retarded time' t' of (40.9). The expressions (40.8) and (40.11) are called the *Lienard-Wiechert potentials*. From them we obtain the electromagnetic fields by differentiation with respect to x, y, z and t. But the Lienard-Wiechert potentials are expressed in terms of $\mathbf{x} = (x, y, z)$ and $t'(\mathbf{x}, t)$; hence we must first find the partial derivatives of t' and then use the chain rule.

PROBLEMS

Problem 40–1. Derive the following formulae for the derivatives of $t'(\mathbf{x}, t)$:

$$\frac{\partial t'}{\partial t} = \frac{1}{1 - \mathbf{n} \cdot \mathbf{v}/c}, \qquad \mathbf{grad}\, t' = -\frac{\mathbf{n}/c}{1 - \mathbf{n} \cdot \mathbf{v}/c}. \qquad (40.12)$$

Problem 40–2. Prove that the fields (38.1) for a moving point charge are:

$$\mathbf{E} = \frac{e}{4\pi\epsilon_0} \frac{1 - v^2/c^2}{r^2(1 - \mathbf{n} \cdot \mathbf{v}/c)^3}(\mathbf{n} - \mathbf{v}/c)$$

$$+\frac{\mu_0}{4\pi} \frac{e}{r(1 - \mathbf{n} \cdot \mathbf{v}/c)^3}\mathbf{n} \times [(\mathbf{n} - \mathbf{v}/c) \times \dot{\mathbf{v}}]$$

$$\mathbf{B} = \frac{1}{c}\mathbf{n} \times \mathbf{E}. \qquad (40.13)$$

We emphasize, again, that all the quantities appearing on the right hand sides of (40.12) and (40.13) are to be taken at the retarded time t' of (40.9).

For an unaccelerated point charge, the second term of $\mathbf{E}$ vanishes, and the fields become

$$\mathbf{E} = \frac{e}{4\pi\epsilon_0} \frac{1 - v^2/c^2}{r^2(1 - \mathbf{n} \cdot \mathbf{v}/c)^3}(\mathbf{n} - \mathbf{v}/c), \qquad \mathbf{B} = \frac{1}{c}\mathbf{n} \times \mathbf{E}. \qquad (40.14)$$

In the rest-frame, they reduce to

$$\mathbf{E} = \frac{e}{4\pi\epsilon_0} \frac{\mathbf{n}}{r^2}, \qquad \mathbf{B} = 0. \qquad (40.15)$$

In fact, (40.14) and (40.15) are connected with each other by the Lorenz transformation formulae (8.7)$_{1,2}$. That they do not follow from each other by applying a Galilei transformation is obvious from the appearance of $1 - v^2/c^2$ in $(40.14)_1$. The Lorenz transformation is superior to the Galilei transformation on the basis of experimental evidence, but this knowledge is not contained in our present calculations. How then does this apparent preference for the Lorenz transformation come about? The reason is that both (40.14) and the electrostatic solution (40.15) assume the aether relations, and we have shown in §8 that every transformation from one aether frame to another must be a Lorenz transformation.

The contribution of the *velocity term* (40.14) to the power $\mathbf{E} \times \mathbf{H} \cdot \mathbf{n}\,dS = r^2 E^2\,d\Omega/(\mu_0 c)$ into the solid angle $d\Omega$ vanishes as r^{-2}. We therefore consider the *acceleration term*

$$\mathbf{E} = \frac{\mu_0}{4\pi} \frac{e}{r(1 - \mathbf{n} \cdot \mathbf{v}/c)^3} \mathbf{n} \times [(\mathbf{n} - \mathbf{v}/c) \times \dot{\mathbf{v}}], \qquad \mathbf{B} = \frac{1}{c}\mathbf{n} \times \mathbf{E}. \quad (40.16)$$

Even by itself, the acceleration term is quite complicated. We shall therefore discuss a few special cases:

(a) *For low velocities* (but arbitrary accelerations),

$$\mathbf{E} \times \mathbf{H} \cdot \mathbf{n}\,dS = \frac{e^2}{4\pi\epsilon_0} \frac{\dot{v}^2 - \dot{v}_n^2}{c^3} \frac{d\Omega}{4\pi} = \frac{e^2 \dot{v}^2}{4\pi\epsilon_0 c^3} \sin^2\theta \frac{d\Omega}{4\pi}, \quad (40.17)$$

where θ is the angle between $\mathbf{n}$ and $\dot{\mathbf{v}}$. The angular distribution of (40.17) is the same as that of electric dipole radiation (cf. (39.14)) or the linear antenna, and is shown in Figure 16.

The total power is obtained by integrating over all angles (cf. (39.15)):

$$P = \frac{1}{4\pi\epsilon_0} \frac{2e^2 \dot{v}^2}{3c^3}. \quad (40.18)$$

This formula, like (39.15) – which it resembles – is also called *Larmor's formula*.

(b) *For the case of an acceleration that is parallel to the velocity,*

$$\mathbf{E} \times \mathbf{H} \cdot \mathbf{n}\,dS = \frac{e^2 \dot{v}^2}{4\pi\epsilon_0 c^3} \frac{\sin^2\theta}{(1 - \beta\cos\theta)^6} \frac{d\Omega}{4\pi}, \quad (40.19)$$

where θ is now the angle between $\mathbf{n}$ and the velocity $\mathbf{v}$, and $\beta = v/c$. This is symmetric about $\mathbf{v}$ and vanishes in the fore and aft directions, but now the denominator causes the lobes of Figure 16 to be displaced towards the forward direction (Figure 17). For relativistic speeds, when $\beta = v/c$ is close to 1, the displacement is quite pronounced, and most of the radiation takes place in the forward direction.

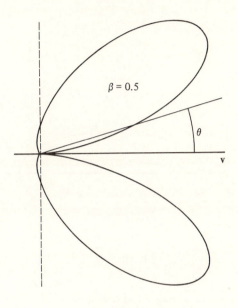

$\beta = 0.5$

θ

$\mathbf{v}$

FIGURE 17

(c) *When the acceleration is normal to the velocity*, as in circular motion,

$$\mathbf{E} \times \mathbf{H} \cdot \mathbf{n} \, dS$$

$$= \frac{e^2 \dot{v}^2}{4\pi\epsilon_0 c^3} \left[\frac{1}{(1 - \beta\cos\theta)^4} - \frac{(1 - \beta^2)\sin^2\theta\cos^2\phi}{(1 - \beta\cos\theta)^6} \right] \frac{d\Omega}{4\pi}, \qquad (40.20)$$

where θ is, as before, the angle between $\mathbf{n}$ and $\mathbf{v}$, and ϕ is the azimuthal angle of $\mathbf{n}$ relative to the plane formed by $\mathbf{v}$ and $\dot{\mathbf{v}}$; specifically, $v_n = v\cos\theta$ and $\dot{v}_n = \dot{v}\sin\theta\cos\phi$. This is symmetric with respect to the latter plane; and in that plane, it vanishes along the two directions for which $\cos\theta = \beta = v/c$. Since the second term in the square brackets is

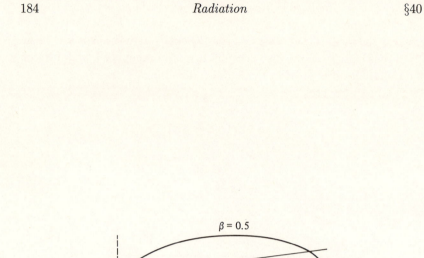

FIGURE 18

positive, the expression is majorized by the first term. The latter has its
maximum in the forward direction (for which the second term vanishes).
The radiation therefore has a maximum in the forward direction. For
relativistic speeds, this becomes a sharp ray. Figure 18 shows the angular
distribution in the plane formed by **v** and **v̇**.

CHAPTER XI

Electromagnetic Wave

Propagation

41. Monochromatic plane waves

In Chapter X we treated the *generation* of electromagnetic waves, which was governed by inhomogeneous wave equations. The *propagation* of electromagnetic waves is characterized by homogeneous wave equations. This does not mean that we are going to set all charges and currents equal to zero, because that would limit us to materials without response. In fact, we shall only assume that the *free* charge density vanishes. This is justifiable in insulators as well as in metals (where charge relaxation causes the charge density to vanish, even at very high frequencies). We shall also confine the discussion to electromagnetic waves that are sufficiently weak for the material response to be linear.† This simplifies matters, because it leads to a linear system of equations. But, as we found in our discussion of the skin effect (§35), the response ceases to be instantaneous at high frequencies, because it cannot keep up with the rapid changes in the electromagnetic field. As a result, Maxwell's equations cease to be partial differential relations. We therefore resolve the fields by a Fourier expansion into monochromatic components, each of which depends on the time through the factor $e^{-i\omega t}$, and replace the instantaneous response relations $\mathcal{J} = \sigma \mathcal{E}$, $\mathbf{P} = \epsilon_0 \chi \mathcal{E}$ and $\mu_0 \mathcal{M} = \chi_B \mathbf{B}$ by the appropriate relations for the monochromatic components (cf. §35). In matter which is stationary (with respect to an aether frame), we are

---

† The theory of waves in which non-linear response is important is called *non-linear optics*.

thus led to the following equations for the monochromatic components of **E** and **H**:

$$\text{div}\,\epsilon\mathbf{E} = 0, \qquad \text{curl}\,\mathbf{H} = (\sigma - i\omega\epsilon)\mathbf{E},$$

$$\text{div}\,\mu\mathbf{H} = 0, \qquad \text{curl}\,\mathbf{E} = i\omega\mu\mathbf{H}. \tag{41.1}$$

In these equations ϵ, σ and μ depend on ω and on the position **x**. In uniform materials they depend only on ω, and the equations become

$$\text{div}\,\mathbf{E} = 0, \qquad \text{div}\,\mathbf{H} = 0,$$

$$\text{curl}\,\mathbf{H} = (\sigma - i\omega\epsilon)\mathbf{E}, \qquad \text{curl}\,\mathbf{E} = i\omega\mu\mathbf{H}. \tag{41.2}$$

Unless $\sigma - i\omega\epsilon$ or $\omega\mu$ happen to vanish, $(41.2)_{1,2}$ follow from $(41.2)_{3,4}$.

In an insulator $\sigma = 0$. Real dielectrics, however, do have finite, even if small, conductivities. It is customary to replace $\sigma - i\omega\epsilon$ by $-i\omega\epsilon$, where the new ϵ is complex. Denoting its real and imaginary parts by ϵ' and ϵ'', we have

$$\epsilon = \epsilon' + i\epsilon'', \qquad \epsilon'' = \sigma(\omega)/\omega, \tag{41.3}$$

and $(41.2)_3$ becomes $\text{curl}\,\mathbf{H} = -i\omega\epsilon\mathbf{E}$. In fact, handbooks on material properties never mention the conductivity of dielectrics. Instead, they list ϵ'', or the angle $\delta = \tan^{-1}(\epsilon''/\epsilon')$, at specified frequencies.

Metals, too, can of course be described by (41.3); for them, the imaginary part of the complex 'permittivity' ϵ is large compared to the real part. Often, however, $\sigma - i\omega\epsilon$ is replaced in discussions of metals by a complex $\sigma = \sigma' + i\sigma''$, where the (normally small) imaginary part σ'' is related to the ordinary dielectric susceptibility. With this notation, $(41.2)_3$ becomes $\text{curl}\,\mathbf{H} = \sigma\mathbf{E}$.

In creating the link between electromagnetism and thermodynamics we introduced (in §12) the product $\mathcal{E} \times \mathcal{H}$ as an extra energy flux. Its contribution to the rate of energy increase $\rho\dot{e}$ in stationary matter (where $\mathcal{E} = \mathbf{E}$ and $\mathcal{H} = \mathbf{H}$) is $-\text{div}\,\mathbf{E} \times \mathbf{H}$ per unit volume (cf. (12.5)). This is called the rate of *absorption* (of radiation per unit volume). Positive absorption does not necessarily mean that the material (at the point considered) is warming up: according to the equation of energy balance (12.5), absorption may be counteracted by cooling (e.g., positive div **q**) or by negative mechanical power.

In order to calculate the rate of absorption in a monochromatic field of frequency ω we use the vector identity $-\operatorname{div} \mathbf{E} \times \mathbf{H} = \mathbf{E} \cdot \operatorname{curl} \mathbf{H} - \mathbf{H} \cdot \operatorname{curl} \mathbf{E}$ and substitute for the curls from $(41.2)_{3,4}$. Using the notation (41.3), we have $\operatorname{curl} \mathbf{H} = -i\omega\epsilon\mathbf{E}$ and $\operatorname{curl} \mathbf{E} = i\omega\mu\mathbf{H}$; we shall (formally) allow for a complex magnetic permeability, $\mu = \mu' + i\mu''$. It is important to notice that, in calculating the absorption, we are forming products of quantities (e.g., $\mathbf{E}$ and $\operatorname{curl} \mathbf{H}$), which are in turn given by the real parts of complex numbers (e.g., $\mathbf{E}$ and $-i\omega\epsilon\mathbf{E}$), and that each one of these fluctuates with a frequency ω. If $a(t)$ and $b(t)$ are complex, and each one is proportional to $e^{-i\omega t}$, the product of their real parts is $(a + a^*)(b + b^*)/4$ (where a^* denotes the complex conjugate of a), which is a sum of four terms. Two of these fluctuate (with a frequency 2ω), and the remaining two are constant. The time average of the product is therefore $(ab^* + ba^*)/4$, which is $\Re(\frac{1}{2}ab^*)$. This result (in which the product was only assumed to satisfy the distributive law of multiplication) applies, in particular, to scalar and vector products of complex vectors. Using it, we find that, in a monochromatic field, the average rate of absorption (per unit volume) is

$$\tfrac{1}{2}\omega\bigl(\epsilon''|E|^2 + \mu''|H|^2\bigr). \tag{41.4}$$

We have seen that a non-zero ϵ'' is to be expected, because it is merely an alternative notation for a non-zero conductivity. It should also be noted that the real part ϵ' of the permittivity behaves quite differently from the static dielectric constant (which corresponds to $\omega = 0$): it is quite common for ϵ' to vanish, and even to become negative in some frequency ranges. The complex permeability μ, on the other hand, has been introduced purely formally. In order to assess the relative roles of magnetization and polarization, we estimate the two terms of the bound current density $\operatorname{curl} \mathbf{M} + \mathbf{P}_t$. If ℓ is the scale over which the fields vary, then $\operatorname{curl} \mathbf{M}$ is of the order of $\chi_B B/(\mu_0 \ell)$, and $\mathbf{P}_t$ is of the order of $\omega\epsilon_0\chi E$, or $\omega^2\epsilon_0\chi\ell B$ (since $\operatorname{curl} \mathbf{E} = -\mathbf{B}_t$). Noting that $c/\omega = \lambda_0/2\pi$, where λ_0 is the wavelength corresponding to the frequency ω in vacuum, we find that the ratio of $\operatorname{curl} \mathbf{M}$ to $\mathbf{P}_t$ is $(\lambda_0/2\pi\ell)^2(\chi_B/\chi)$. In macroscopic samples, at optical frequencies, the first factor is quite small. As for the second factor, we have no reason to suppose that χ_B at high frequencies is significantly different from the tiny, static values that it has in para- and diamagnets (less than 10^{-4}).† Compared with polarization, magnetization is therefore quite insignificant in optical phenomena. In what follows, we shall take μ to be real and close to μ_0. The rate of absorption will then reduce to the first term of (41.4), which in view of $(41.3)_2$ is just the average Ohmic loss (cf. §32). A medium in which ϵ'' vanishes is called *transparent*.

We shall now consider solutions of (41.2) that have the dependence $e^{i\mathbf{k}\cdot\mathbf{x}}$ on the coordinates $\mathbf{x} = (x, y, z)$. For any vector field $\mathbf{f}$ with this spatial dependence, div $\mathbf{f} = i\mathbf{k} \cdot \mathbf{f}$ and **curl** $\mathbf{f} = i\mathbf{k} \times \mathbf{f}$. The equations $(41.2)_{3,4}$ (with the notation (41.3)) then become

$$\mathbf{k} \times \mathbf{H} = -\omega\epsilon\mathbf{E}, \qquad \mathbf{k} \times \mathbf{E} = \omega\mu\mathbf{H}. \tag{41.5}$$

Solutions $\mathbf{E}$ and $\mathbf{H}$ of this linear, homogeneous system will vanish unless

$$k^2 = \omega^2\epsilon\mu. \tag{41.6}$$

In vacuum, where $\sigma = 0$, $\epsilon = \epsilon_0$ and $\mu = \mu_0$, $\mathbf{k}$ is a real vector, but in a medium this need not be so, and $\mathbf{k}$ may be complex. As usual, we shall write $\mathbf{k} = \mathbf{k}' + i\mathbf{k}''$. Then (41.6) becomes

$$k'^2 - k''^2 + 2i\mathbf{k}' \cdot \mathbf{k}'' = \omega^2(\epsilon' + i\epsilon'')\mu. \tag{41.7}$$

This may be solved for the real and imaginary parts of $\mathbf{k}$. If $\epsilon'' = 0$ (so that the medium is transparent) and $\epsilon'\mu > 0$, $\mathbf{k}$ is real and has the magnitude

$$k = \omega\sqrt{\epsilon\mu} = \frac{n\omega}{c}, \tag{41.8}$$

where $n = c\sqrt{\epsilon\mu}$ is the *refractive index* of the medium. According to (41.5), the electric and magnetic fields are in this case in a plane normal to $\mathbf{k}$ and are also perpendicular to each other. They form a plane wave with the dependence $e^{i(\mathbf{k}\cdot\mathbf{x}-\omega t)}$. And this wave has the phase velocity $\omega/k = c/n$.

If $\mathbf{k}$ is complex, $i\mathbf{k} \cdot \mathbf{x} = i\mathbf{k}' \cdot \mathbf{x} - \mathbf{k}'' \cdot \mathbf{x}$, and we see from the spatial dependence $e^{i\mathbf{k}'\cdot\mathbf{x}}e^{-\mathbf{k}''\cdot\mathbf{x}}$ that the planes of constant phase are normal to $\mathbf{k}'$, whereas those of constant amplitude are normal to $\mathbf{k}''$. The fields

---

† We are concerned with weak electromagnetic fields; a weak, variable field is one of the common methods of demagnetizing ferromagnets, thus reducing their μ to $\approx \mu_0$.

themselves are, in general, constant on neither. Such solutions are not really plane waves; they are called *inhomogeneous plane waves*.

In the special case in which $\mathbf{k}'$ and $\mathbf{k}''$ are parallel, the solution *is* a plane wave, although a damped one. If $\mathbf{l}$ is a unit vector in the common direction of $\mathbf{k}'$ and $\mathbf{k}''$, we have $\mathbf{k} = k\mathbf{l}$, where k satisfies (41.6). We denote

$$k = (n + i\kappa)\frac{\omega}{c}, \tag{41.9}$$

with real n and κ. They are, respectively, called the *refractive index* and the *absorption coefficient* of the medium. With this notation, (41.7) becomes

$$n^2 - \kappa^2 + 2in\kappa = c^2(\epsilon' + i\epsilon'')\mu. \tag{41.10}$$

PROBLEM

Problem 41–1. Show that the solution of (41.10) is

$$n = c\sqrt{\tfrac{1}{2}\mu}\sqrt{\epsilon' + \sqrt{\epsilon'^2 + \epsilon''^2}},$$

$$\kappa = c\sqrt{\tfrac{1}{2}\mu}\sqrt{-\epsilon' + \sqrt{\epsilon'^2 + \epsilon''^2}}. \tag{41.11}$$

The wave, which has the dependence $e^{i(nx-ct)\omega/c}e^{-\kappa x\omega/c}$, is damped if $\kappa > 0$. According to (41.10), this will always be the case in an absorbing medium, because $\kappa \neq 0$ if $\epsilon'' \neq 0$. In metals, where ϵ' is small compared to $\epsilon'' = \sigma/\omega$,

$$n = \kappa = c\sqrt{\frac{\mu\sigma}{2\omega}} = \frac{c}{\omega\delta}, \tag{41.12}$$

where δ is the skin depth (cf. $(35.7)_3$). Damping can also occur, *without* any absorption of radiation, in a transparent medium: if $\epsilon'' = 0$ and ϵ' is negative, (41.10) gives $n = 0$ and $\kappa = c\sqrt{-\epsilon'\mu}$.

PROBLEM

Problem 41–2. Prove that, in the case of damping in a transparent medium, the time average of the Poynting vector $\mathbf{E} \times \mathbf{H}$ vanishes.

42. Reflection and refraction

In a transparent medium with $\epsilon' > 0$, according to the results of the last section, a plane monochromatic wave which is propagating in the direction $\mathbf{l}$ is characterized by

$$\mathbf{k} = k\mathbf{l}, \qquad k = \omega\sqrt{\epsilon\mu} = \frac{n\omega}{c},$$

$$\mathbf{H} = \sqrt{\frac{\epsilon}{\mu}}\mathbf{l} \times \mathbf{E}, \tag{42.1}$$

and the fields have the dependence $e^{i(\mathbf{k}\cdot\mathbf{x}-\omega t)}$ on the coordinates. Consider now such a wave which impinges normally on the plane boundary between two different media, both of which are at rest (with respect to an aether frame). Since $\mathbf{l}$ is normal to the boundary, $\mathbf{E}$ and $\mathbf{H}$ in the incident wave are both parallel to the boundary. Both must therefore be continuous. Can the wave then simply proceed, without any change, through the interface into the second medium? Clearly not, because $(42.1)_3$ cannot be satisfied with the same fields if ϵ/μ is discontinuous. We must therefore conclude that, if anything does go through into the second medium, there must be a corresponding disturbance in the first medium. In other words, *transmission* must result in *reflection*. Is it possible (conversely) for an incident wave to be reflected without anything passing through? Obviously yes, and this situation constitutes a perfect mirror.

We now turn to the more general case of oblique incidence from a transparent medium (designated 1) into a second medium (designated 2), which is not necessarily transparent. We denote the propagation vectors and frequencies of the incident, transmitted and reflected waves (respectively) by $\mathbf{k}^{(i)}$, $\mathbf{k}^{(t)}$, $\mathbf{k}^{(r)}$, $\omega^{(i)}$, $\omega^{(t)}$, $\omega^{(r)}$, and choose the coordinates so that the (stationary) surface of separation is the plane $z = 0$. On $z = 0$, the x and y components of $\mathbf{E}$ and $\mathbf{H}$ must be continuous. Since the fields have the dependence $e^{i(\mathbf{k}\cdot\mathbf{x}-\omega t)}$, these conditions can only be satisfied if the three arguments $\mathbf{k} \cdot \mathbf{x} - \omega t$ are all equal there:

$$xk_x^{(r)} + yk_y^{(r)} - \omega^{(r)}t = xk_x^{(i)} + yk_y^{(i)} - \omega^{(i)}t,$$

$$xk_x^{(t)} + yk_y^{(t)} - \omega^{(t)}t = xk_x^{(i)} + yk_y^{(i)} - \omega^{(i)}t. \tag{42.2}$$

These equations must hold all over the boundary at all times, that is, for all x, y and t. It follows, first, that the three frequencies must be equal.† Similarly, we must have $k_x^{(t)} = k_x^{(r)} = k_x^{(i)}$ and $k_y^{(t)} = k_y^{(r)} = k_y^{(i)}$. We define the *plane of incidence* as the plane that contains the (real) propagation vector $\mathbf{k}^{(i)}$ of the incident wave and the normal to the surface of separation, and choose the coordinates so that this is the xz plane. By this choice, $k_y^{(i)} = 0$. It follows that $k_y^{(t)}$ and $k_y^{(r)}$, too, must vanish. This means that all three waves have their propagation vectors in the same plane, and we have the situation shown in Figure 19.

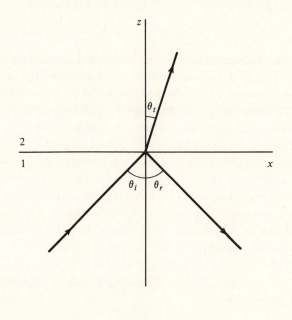

FIGURE 19

We have not yet made any use of the equality of the x components of

† That is because the surface of separation has been assumed stationary. If it were moving, we would have to apply the jump conditions on $z(t)$, rather than on $z = 0$; the resulting relations (called *Doppler relations*) between the ω's would then involve the normal speed of the surface.

the **k**'s. In the transparent medium 1, we have (cf. Figure 19)

$$k_x^{(i)} = (\omega/c)n_1 \sin\theta_i, \qquad k_x^{(r)} = (\omega/c)n_1 \sin\theta_r, \qquad (42.3)$$

where $n_1 = c\sqrt{\epsilon_1\mu_1}$. The equality $k_x^{(r)} = k_x^{(i)}$ gives the *law of reflection*

$$\theta_r = \theta_i. \qquad (42.4)$$

If the second medium, too, is transparent,

$$k_x^{(t)} = (\omega/c)n_2 \sin\theta_t, \qquad (42.5)$$

and the equality $k_x^{(t)} = k_x^{(i)}$ gives *Snell's law of refraction*

$$n_2 \sin\theta_t = n_1 \sin\theta_i. \qquad (42.6)$$

So far, our discussion has merely concerned itself with directions. In order to deal with the amplitudes, we must apply the jump conditions, which require the continuity of the x and y components of **E** and $\mathbf{H} = \mathbf{k}\times\mathbf{E}/(\omega\mu)$ across $z = 0$. Since the k_y's all vanish, the jump conditions are seen to fall into two groups: one, which involves E_y (the component perpendicular to the plane of incidence) and $H_x = -k_z E_y/(\omega\mu)$; and the other, involving E_x and E_z (the components in the plane of incidence) and $H_y = (k_z E_x - k_x E_z)/(\omega\mu)$. The plane formed by the propagation and electric vectors is called the *plane of polarization* of the wave. The foregoing two groups are therefore waves polarized perpendicular and parallel to the plane of incidence. In order to implement the jump conditions we need the z components of the **k**'s . Since the k_y's all vanish, $k_z^2 = k^2 - k_x^2 = \omega^2\epsilon\mu - k_x^2$, and we obtain k_z by taking the square root with the appropriate sign. According to (42.3), together with the equality of the k_x's, we have

$$k_z^{(i)} = \omega\sqrt{\epsilon_1\mu_1}\cos\theta_i = (\omega/c)n_1\cos\theta_i,$$

$$k_z^{(r)} = -\omega\sqrt{\epsilon_1\mu_1}\cos\theta_i = -(\omega/c)n_1\cos\theta_i,$$

$$k_z^{(t)} = \omega\sqrt{\epsilon_2\mu_2 - \epsilon_1\mu_1\sin^2\theta_i} = (\omega/c)\sqrt{c^2\epsilon_2\mu_2 - n_1^2\sin^2\theta_i}. \qquad (42.7)$$

In the following formulae we shall set $\mu_1 = \mu_2 = \mu_0$. Taking, first, the case in which the electric field in the incident wave is perpendicular to the plane of incidence, we have $E_y^{(i)} + E_y^{(r)} = E_y^{(t)}$ and $k_z^{(i)}E_y^{(i)} + k_z^{(r)}E_y^{(r)} =$

$k_z^{(t)} E_y^{(t)}$. The solution of these equations gives *Fresnel's* formulae (we omit the subscript y):

$$E^{(r)} = \frac{n_1 \cos\theta_i - \sqrt{\epsilon_2/\epsilon_0 - n_1^2 \sin^2\theta_i}}{n_1 \cos\theta_i + \sqrt{\epsilon_2/\epsilon_0 - n_1^2 \sin^2\theta_i}} E^{(i)},$$

$$E^{(t)} = \frac{2n_1 \cos\theta_i}{n_1 \cos\theta_i + \sqrt{\epsilon_2/\epsilon_0 - n_1^2 \sin^2\theta_i}} E^{(i)}. \tag{42.8}$$

If the second medium, too, is transparent, we can make use of Snell's law $n_1 \sin\theta_i = n_2 \sin\theta_t$. Fresnel's formulae then become

$$E^{(r)} = \frac{\sin(\theta_t - \theta_i)}{\sin(\theta_t + \theta_i)} E^{(i)}, \qquad E^{(t)} = \frac{2\cos\theta_i \sin\theta_t}{\sin(\theta_t + \theta_i)} E^{(i)}. \tag{42.9}$$

A similar calculation can be carried out for the case in which the electric vector in the incident wave is *in* the plane of incidence. The magnetic vector is then perpendicular to this plane, and Fresnel's formulae (in terms of $H = H_y$) are

$$H^{(r)} = \frac{(\epsilon_2/\epsilon_0)\cos\theta_i - n_1\sqrt{\epsilon_2/\epsilon_0 - n_1^2 \sin^2\theta_i}}{(\epsilon_2/\epsilon_0)\cos\theta_i + n_1\sqrt{\epsilon_2/\epsilon_0 - n_1^2 \sin^2\theta_i}} H^{(i)},$$

$$H^{(t)} = \frac{2(\epsilon_2/\epsilon_0)\cos\theta_i}{(\epsilon_2/\epsilon_0)\cos\theta_i + n_1\sqrt{\epsilon_2/\epsilon_0 - n_1^2 \sin^2\theta_i}} H^{(i)}. \tag{42.10}$$

If the second medium, too, is transparent, these formulae become

$$H^{(r)} = \frac{\tan(\theta_i - \theta_t)}{\tan(\theta_i + \theta_t)} H^{(i)},$$

$$H^{(t)} = \frac{\sin 2\theta_i}{\sin(\theta_t + \theta_i)\cos(\theta_i - \theta_t)} H^{(i)}. \tag{42.11}$$

Each of the waves is associated with an energy flux, the time average of which is $\Re(\frac{1}{2}\mathbf{E} \times \mathbf{H}^*)$. In order to obtain the z component, which gives the amount of energy crossing unit area of the surface of separation per unit time, we must multiply by $\cos\theta$. The ratio R of the normal components of the average energy flux in the reflected and incident waves is called the *reflectivity*. Similarly, the corresponding ratio T of the transmitted and incident waves is called the *transmissivity*.

We shall first calculate R and T (from (42.9) and (42.11)) for the case in which both media are transparent. For an incident wave with the electric vector perpendicular to the plane of incidence (we indicate this by the subscript $\perp$) we have

$$R_\perp = \frac{\sin^2(\theta_i - \theta_t)}{\sin^2(\theta_i + \theta_t)}, \qquad T_\perp = \frac{\sin 2\theta_i \sin 2\theta_t}{\sin^2(\theta_i + \theta_t)}. \qquad (42.12)$$

We note that $R_\perp + T_\perp = 1$, as expected. For an incident wave with the electric vector in the plane of incidence (subscript $\|$), we have

$$R_\| = \frac{\tan^2(\theta_i - \theta_t)}{\tan^2(\theta_i + \theta_t)}, \qquad T_\| = \frac{\sin 2\theta_i \sin 2\theta_t}{\sin^2(\theta_i + \theta_t) \cos^2(\theta_i - \theta_t)}; \qquad (42.13)$$

again, $R_\| + T_\| = 1$.

For normal incidence the distinction between (42.12) and (42.13) disappears, and either of them reduces to

$$R = \frac{(n-1)^2}{(n+1)^2}, \qquad T = \frac{4n}{(n+1)^2}, \qquad (42.14)$$

where $n = n_2/n_1$. In all cases, when n_2 approaches n_1, the reflectivity goes to zero and the transmissivity goes to 1.

In the special case when $\theta_i + \theta_t = \pi/2$, that is, when the transmitted and reflected rays are perpendicular to each other, $\tan(\theta_i + \theta_t)$ becomes infinite. According to (42.13)$_1$, $R_\|$ vanishes in this case. The angle of incidence for which this special case arises is given by

$$\tan \theta_i = n = \frac{n_2}{n_1}, \qquad (42.15)$$

which follows from $\sin \theta_t = \sin(\pi/2 - \theta_i) = \cos \theta_i$, together with Snell's law (42.6). This θ_i is called the *Brewster angle*. Thus, when light is incident under this angle, the electric vector in the reflected wave will have no component in the plane of incidence. Such a reflected wave is said to be *totally polarized*; the Brewster angle is therefore also called the (incidence) *angle of total polarization*.

It may be noted that, according to (42.12)$_1$ and (42.13)$_1$, $R_\| < R_\perp$ (except when $\theta_i = 0$ or $\pi/2$). Natural light (an equal mixture of waves with $\mathbf{E}$ in every direction perpendicular to the propagation vector) is therefore always partially polarized by reflection: the reflected wave will have its electric vector predominantly normal to the plane of incidence.

If $n_2 > n_1$, the second medium is said to be optically denser than the first medium. In this case, according to Snell's law (42.6), $\sin\theta_t < \sin\theta_i$; hence $\theta_t < \theta_i$. If, on the other hand, the second medium is optically less dense than the first, $\sin\theta_t$ becomes unity (so that the refracted wave is propagated along the surface of separation) when

$$\sin\theta_i = n = \frac{n_2}{n_1}. \tag{42.16}$$

This angle of incidence is called the angle of *total reflection*.

PROBLEM

Problem 42–1. Show that $R_\parallel = R_\perp = 1$ at the angle of total reflection.

If (for $n_2 < n_1$) θ_i exceeds the angle of total reflection, $k_z^{(t)}$ becomes imaginary (cf. (42.7)), and the fields in the second medium undergo damping. Since the second medium is transparent, this damping is not accompanied by absorption. It is easily checked that the Poynting vector, though not zero, has a vanishing time average. Energy therefore flows to and fro, but there is no lasting flow. A modern application of total reflection is found in *optical fibres*. These are tubes made of highly transparent material (which is optically denser than the surroundings). Electromagnetic waves with propagation vectors making a small angle with the axis of the tube are then bounced off the walls by total reflection, that is, with no loss of energy.

The foregoing discussion of reflectivity and transmissivity has been confined to the case in which both media are transparent. If the second medium is not transparent, R and T must be calculated from (42.8) and (42.10). The general formulae are rather unwieldy. For normal incidence ($\theta_i = 0$), however, the real and imaginary parts of the complex $\mathbf{k}^{(t)}$ are both in the z direction. The wave in medium 2 then becomes homogeneous, and we can use (41.9).

PROBLEM

Problem 42–2. Show that, for normal incidence,

$$R = \frac{(n-1)^2 + \kappa^2}{(n+1)^2 + \kappa^2}, \tag{42.17}$$

where $n = n_2/n_1$ and $\kappa = \kappa_2/\kappa_1$.

We have seen that, in metals, n and κ are of the order of the wavelength, divided by the skin depth (cf. (41.12)). So long as the frequency is not too high, n^2 and κ^2 are therefore large, and $R \approx 1 - 2/n$ is close to 1. At optical frequencies the simple formulae (41.12) break down, because $\omega\epsilon$ is no longer small compared to σ, but n^2 and κ^2 are usually still large. Metals are therefore good reflectors, even at optical frequencies.

43. Conditions at metallic surfaces

We shall now examine in more detail the case in which the second medium is a metal, with a conductivity σ that is large compared to $\omega\epsilon$. In this case we have (cf. (35.4))

$$\mathbf{curl\ H} = \sigma\mathbf{E}, \qquad \mathbf{curl\ E} = i\omega\mathbf{B}. \tag{43.1}$$

In a *perfect conductor* ($\sigma \to \infty$), $\mathbf{E} = \mathbf{curl\ H}/\sigma$ must vanish. Along with $\mathbf{E}$, any variable magnetic field ($\omega \neq 0$) must vanish as well (according to (43.1)$_2$). From the jump conditions $\mathbf{n} \times [\mathbf{E}] = 0$ and $\mathbf{n} \cdot [\mathbf{B}] = 0$ it then follows that, just outside the metal, the following boundary conditions hold:

$$\mathbf{n} \times \mathbf{E} = 0, \qquad B_n = 0. \tag{43.2}$$

The remaining jump conditions $\mathbf{n} \cdot [\mathbf{D}] = \tau$ (where τ now denotes the surface charge density, because σ is the conductivity) and $\mathbf{n} \times [\mathbf{H}] = \mathbf{K}$ determine the surface charge and current densities on the walls:

$$\tau = -D_n, \qquad \mathbf{K} = \mathbf{H} \times \mathbf{n}. \tag{43.3}$$

In these formulae $\mathbf{n}$ points *into* the metal.

PROBLEM

Problem 43–1. Prove that the stress on a perfectly conducting wall is

$$Tn = \tfrac{1}{2}(\tau \mathbf{E} + \mathbf{K} \times \mathbf{B}).\tag{43.4}$$

This is a normal stress. The first term is a tension, and the second a pressure (called *radiation pressure*).

The boundary conditions (43.2) and the formulae (43.3) were obtained in the limit $1/\sigma \to 0$ of a perfect conductor. We shall now seek the next order terms, assuming that $1/\sigma$, though small, does not vanish; they will actually turn out to be of order $1/\sqrt{\sigma}$. As in §42, we consider the case of a monochromatic wave that impinges on a metal from a transparent medium. In the metal, the equations (43.1) become

$$i\mathbf{k} \times \mathbf{H} = \sigma \mathbf{E}, \qquad \mathbf{k} \times \mathbf{E} = \omega\mu\mathbf{H}.\tag{43.5}$$

It follows that $(\mathbf{k}' + i\mathbf{k}'')^2 = i\mu\sigma\omega$, or

$$k'^2 - k''^2 + 2i\mathbf{k}' \cdot \mathbf{k}'' = i\mu\sigma\omega;\tag{43.6}$$

Taking real and imaginary parts, we have

$$k'^2 = k''^2, \qquad \mathbf{k}' \cdot \mathbf{k}'' = \frac{\mu\sigma\omega}{2} = \frac{1}{\delta^2},\tag{43.7}$$

where δ is the skin depth (cf. $(35.7)_3$). From the relations (42.2) we obtain, as before, $k_x = k'_x + ik''_x = k_x^{(i)}$ and $k_y = k'_y + ik''_y = 0$. Substitution in (43.6) gives

$$k_x^{(i)\,2} + k_z'^{\,2} = k_z''^{\,2}, \qquad k_z'k_z'' = \frac{1}{\delta^2}.\tag{43.8}$$

These can be solved for k'_z and k''_z. By $(43.8)_2$, k'_z and k''_z have the same sign, which means that the wave decays as it propagates. According to $(42.3)_1$, $k_x^{(i)}$ is at most equal to the reciprocal wavelength $(\omega/c)n_1$ in medium 1. We shall assume this to be small compared to $1/\delta$. In this approximation (of small δ), the solution of (43.8) gives $k'_z = k''_z = 1/\delta$, so that the fields inside the metal have the dependence

$$e^{-z/\delta}e^{i[x(\omega/c)n_1 \sin \theta_i + z/\delta - \omega t]}.\tag{43.9}$$

The wave inside the metal is therefore inhomogeneous: the constant amplitude surfaces are parallel to the metal surface, but the constant phase surfaces are not. In the limit of small δ we may neglect this inhomogeneity and take $\mathbf{k}$ to be in the z direction $\mathbf{n}$:

$$\mathbf{k} = \frac{1+i}{\delta}\mathbf{n}. \tag{43.10}$$

Substituting this in $(43.5)_1$, we obtain

$$\mathbf{E} = \frac{i-1}{\sigma\delta}\mathbf{n} \times \mathbf{H}. \tag{43.11}$$

Since $(\sigma\delta)^{-1} \propto \delta$, this shows that $\mathbf{E}$ is small and (in this approximation, predominantly) parallel to the surface of the metal. Just inside the metal (at $z = 0$), equation (43.11) is a relation between the tangential components of $\mathbf{E}$ and $\mathbf{H}$. But these are both continuous (in the approximation of $(43.1)_1$, with finite σ, $\mathbf{n} \times \mathbf{H}$ is continuous). Hence the fields just outside the metal must obey the same relation:

$$\mathbf{E}_s = \frac{i-1}{\sigma\delta}\mathbf{n} \times \mathbf{H}_s. \tag{43.12}$$

Thus E_s is of order $H_s\delta$. The normal component E_n outside the metal is *not* constrained by these considerations. Nor is it continuous. According to $(42.1)_3$ it is of the same order as H_s.

The normal component B_n is continuous. Inside the metal, according to $(43.5)_2$, it is $(\mathbf{k} \times \mathbf{E})_n/\omega$, which involves the tangential components of $\mathbf{k}$ and $\mathbf{E}$. Since the tangential component of $\mathbf{k}$ is $k_x^{(i)}$, B_n is of the order of E_s, or $H_s\delta$.

By (43.12), the time averaged flux into the metal is

$$\Re(\tfrac{1}{2}\mathbf{E} \times \mathbf{H}^*) = \frac{|H_s|^2}{2\sigma\delta}\mathbf{n}. \tag{43.13}$$

PROBLEMS

Problem 43–2. Show that, in the metal,

$$\int_0^\infty \Re\left(\frac{\mathbf{j} \cdot \mathbf{j}^*}{2\sigma}\right) dz = \frac{|H_s|^2}{2\sigma\delta}. \tag{43.14}$$

The time averaged flow of radiation into a thick slab of metal is therefore equal to the time averaged Joule warming. In a thin slab (of thickness $\leq \delta$) some of the radiation will leave through the other side.

Problem 43-3. Show that, in the metal,

$$\int_0^\infty \mathbf{j}\, dz = \mathbf{H} \times \mathbf{n}. \qquad (43.15)$$

According to (43.9), $\mathbf{j}$ is confined to a few δ near the surface. When δ is small, the left hand side effectively becomes a surface current density.

Since E_s/E_n and H_n/H_s are each of order δ, we obtain the ideal conditions (43.2) in the limit $\delta \to 0$. The relation (43.15) reduces in this limit to $(43.3)_2$.

44. Waveguides

A waveguide is a long pipe with constant cross section. We shall assume it to be hollow, with perfectly conducting walls, and seek solutions that represent waves propagating along the waveguide. The fact that we can see through a pipe shows that such solutions must exist.

We choose the z axis as the axis of the waveguide. Monochromatic waves that propagate along the pipe will depend on z and t through $e^{i(k_z z - \omega t)}$. For such fields the equations

$$\mathbf{curl\ H} = -i\omega\epsilon_0\mathbf{E}, \qquad \mathbf{curl\ E} = i\omega\mu_0\mathbf{H}, \qquad (44.1)$$

written out in full, are

$$\frac{\partial H_z}{\partial y} - ik_z H_y = -i\omega\epsilon_0 E_x, \qquad \frac{\partial E_z}{\partial y} - ik_z E_y = i\omega\mu_0 H_x,$$

$$ik_z H_x - \frac{\partial H_z}{\partial x} = -i\omega\epsilon_0 E_y, \qquad ik_z E_x - \frac{\partial E_z}{\partial x} = i\omega\mu_0 H_y,$$

$$\frac{\partial H_y}{\partial x} - \frac{\partial H_x}{\partial y} = -i\omega\epsilon_0 E_z, \qquad \frac{\partial E_y}{\partial x} - \frac{\partial E_x}{\partial y} = i\omega\mu_0 H_z. \qquad (44.2)$$

Waves that resemble plane waves propagating in the z direction will have E_z and H_z both equal to zero; they are called *transverse electromagnetic* or, briefly, *TEM* waves. Strictly plane waves (with $\mathbf{E}$ and $\mathbf{H}$

independent of x or y) are, of course, impossible in a waveguide, because they will not satisfy the boundary conditions on the walls. For TEM waves, $(44.2)_{1,2,3,4}$ give

$$\omega^2 = c^2 k_z^2, \qquad \mathbf{E} \cdot \mathbf{H} = 0, \qquad (44.3)$$

as in plane waves. If we use $(44.3)_2$ in $(44.2)_5$, the last two members of (44.2) become

$$\frac{\partial E_x}{\partial x} + \frac{\partial E_y}{\partial y} = 0, \qquad \frac{\partial E_y}{\partial x} - \frac{\partial E_x}{\partial y} = 0; \qquad (44.4)$$

these equations are also satisfied by the transverse components of $\mathbf{H}$. From $(44.4)_2$ we deduce the existence of a potential $V(x, y)$ such that $E_x = -\partial V/\partial x$ and $E_y = -\partial V/\partial y$, or $\mathbf{E} = -\mathbf{grad}_2 V$, where $\mathbf{grad}_2$ is the two-dimensional gradient, defined over the cross section of the waveguide. The first member of (44.4) now gives $\Delta_2 V = 0$, where Δ_2 is the two-dimensional Laplacian. Thus

$$\mathbf{E} = -\mathbf{grad}_2 V, \qquad \Delta_2 V = 0. \qquad (44.5)$$

The magnetic field satisfies a similar pair of equations, with a different potential U (say); according to $(44.3)_2$, the equipotential surfaces of V must be orthogonal to those of U. The boundary condition $E_s = 0$ requires V to be constant on the circumference of the cross section. Similarly, the boundary condition $H_n = 0$ requires $\partial U/\partial n$ to vanish on the circumference of the cross section.

TEM waves are therefore obtained by solving the two-dimensional potential problem over the cross section, and we may use the various methods we have developed in electrostatics for such problems. It is clear that, if the cross section is simply connected, V and U must be constant everywhere. TEM waves can therefore only propagate in waveguides with multiply connected cross sections; for example, in the space between two pipes, one of which lies inside the other, or in the space around two parallel pipes; for only then can V be equal to different constants on different parts of the circumference. The problem of TEM waves in a waveguide is essentially the two-dimensional problem of capacitance (or resistance).

Let us now turn to a different set of solutions by assuming that only one of the fields, $\mathbf{H}$ or $\mathbf{E}$, is transverse. Such solutions are called *transverse magnetic* (TM) and *transverse electric* (TE). In a TM wave, H_z vanishes (but E_z does not), and (44.2) become

$$ik_z H_y = i\omega\epsilon_0 E_x, \qquad \frac{\partial E_z}{\partial y} - ik_z E_y = i\omega\mu_0 H_x,$$

$$ik_z H_x = -i\omega\epsilon_0 E_y, \qquad ik_z E_x - \frac{\partial E_z}{\partial x} = i\omega\mu_0 H_y,$$

$$\frac{\partial H_y}{\partial x} - \frac{\partial H_x}{\partial y} = -i\omega\epsilon_0 E_z, \qquad \frac{\partial E_y}{\partial x} - \frac{\partial E_x}{\partial y} = 0. \qquad (44.6)$$

The first four members give

$$E_x = \frac{ik_z}{\kappa^2}\frac{\partial E_z}{\partial x}, \qquad E_y = \frac{ik_z}{\kappa^2}\frac{\partial E_z}{\partial y},$$

$$H_x = -\frac{i\omega}{\kappa^2}\epsilon_0\frac{\partial E_z}{\partial y}, \qquad H_y = \frac{i\omega}{\kappa^2}\epsilon_0\frac{\partial E_z}{\partial x}, \qquad (44.7)$$

where $\kappa^2 = \omega^2/c^2 - k_z^2$. It is seen that the x and y components of $\mathbf{E}$ and $\mathbf{H}$ are determined by the x and y derivatives of E_z. The latter are the components of the two-dimensional gradient $\mathbf{grad}_2\, E_z$ of E_z. The sixth member of (44.6) is identically satisfied by the $\mathbf{E}_2$ of $(44.7)_{1,2}$. The remaining equation $(44.6)_5$ gives

$$\Delta_2 E_z + \kappa^2 E_z = 0. \qquad (44.8)$$

The boundary condition $\mathbf{E}_s = 0$ requires E_z to vanish on the walls; in particular, on the circumference of a cross section. It is easy to check (with the aid of (44.7)) that the single requirement $E_z = 0$ ensures the vanishing of both $\mathbf{E}_s$ and H_n along the walls.

For TE waves, in which E_z vanishes (but H_z does not), we obtain, instead of (44.7), the equations

$$H_x = \frac{ik_z}{\kappa^2}\frac{\partial H_z}{\partial x}, \qquad H_y = \frac{ik_z}{\kappa^2}\frac{\partial H_z}{\partial y},$$

$$E_x = \frac{i\omega}{\kappa^2}\mu_0\frac{\partial H_z}{\partial y}, \qquad E_y = -\frac{i\omega}{\kappa^2}\mu_0\frac{\partial H_z}{\partial x}. \qquad (44.9)$$

Instead of (44.8), we have

$$\Delta_2 H_z + \kappa^2 H_z = 0. \tag{44.10}$$

The boundary condition $H_n = 0$ now requires $\partial H_z/\partial n$ to vanish on the circumference of the cross section. And this single condition ensures that TE waves will satisfy both $\mathbf{E}_s = 0$ and $H_n = 0$ on the walls.

Both TM and TE waves are therefore obtained by solving the characteristic value problem $-\Delta_2 f = \kappa^2 f$ for the two-dimensional Laplacian operator on the cross section of the waveguide. The TM waves are given by modes that satisfy the boundary condition $f = 0$ on the circumference; the TE waves, by modes that satisfy $\partial f/\partial n = 0$. Unlike TEM waves, TM and TE waves can be propagated in waveguides with simply connected cross sections.

In a waveguide which is filled with a transparent material, (44.1) must be replaced by

$$\mathbf{curl\ H} = -i\omega\epsilon\mathbf{E}, \qquad \mathbf{curl\ E} = i\omega\mu\mathbf{H}. \tag{44.11}$$

It is easily seen that solutions of this system are obtained from the solutions of (44.1) by the replacements

$$\mathbf{E} \to \sqrt{\frac{\epsilon}{\epsilon_0}}\mathbf{E}, \qquad \mathbf{H} \to \sqrt{\frac{\mu}{\mu_0}}\mathbf{H}, \qquad \omega \to \omega\sqrt{\frac{\epsilon\mu}{\epsilon_0\mu_0}}. \tag{44.12}$$

PROBLEMS

Problem 44–1. By using the identity

$$\mathrm{div}_2(f\,\mathbf{grad}_2\,g) = f\Delta_2 g + \mathbf{grad}_2\,f\cdot\mathbf{grad}_2\,g,$$

prove that the TM modes, or the TE modes, corresponding to different eigenvalues κ_m and κ_n are orthogonal in the sense that

$$\int \mathbf{grad}_2\,f_m\cdot\mathbf{grad}_2\,f_n\,dA = 0, \tag{44.13}$$

the integration being over the area of the cross section.

Problem 44–2. With the notation of Problem 44–1, show that

$$\kappa_n^2 = \frac{\int |\,\mathbf{grad}_2\,f_n|^2\,dA}{\int |f_n|^2\,dA}. \tag{44.14}$$

The last result shows that the κ_n^2 are all positive, the smallest one being of the order of ℓ^{-2}, where ℓ is the diameter of the cross section. For each of the κ_n^2, there is a relation

$$\omega^2 = c^2(k_z^2 + \kappa^2) \tag{44.15}$$

between the frequency ω and the wave number k_z of the wave. These waves have the group velocity

$$\frac{\partial \omega}{\partial k_z} = \frac{c k_z}{\sqrt{k_z^2 + \kappa^2}} = \frac{c^2 k_z}{\omega}, \tag{44.16}$$

which varies from 0 (for $k_z = 0$) to c (for $k_z \to \infty$).

Equation (44.15) shows that, for each cross section, there is a frequency $\omega_{min} = c\kappa_{min}$ below which no waves can be propagated in a waveguide with a simply connected cross section. This *cut-off frequency* is of the order of c/ℓ, where ℓ is the diameter of the cross section. It explains, for example, why low-frequency radio stations cannot be received while driving through a tunnel (metal-reinforced concrete walls provide an approximation to metallic walls).

PROBLEMS

Problem 44–3. A waveguide with a rectangular cross section has sides a_x and a_y. Show that the eigenmodes of $-\Delta_2$ for such a rectangle are

$$TM_{n_x n_y} : \qquad \sin \frac{n_x \pi x}{a_x} \sin \frac{n_y \pi y}{a_y},$$

$$TE_{n_x n_y} : \qquad \cos \frac{n_x \pi x}{a_x} \cos \frac{n_y \pi y}{a_y}, \tag{44.17}$$

where n_x and n_y are integers: both of them non-zero for the TM modes, and at least one of them non-zero for the TE modes. Show that, for either type,

$$\kappa^2 = \pi^2 \left(\frac{n_x^2}{a_x^2} + \frac{n_y^2}{a_y^2} \right). \tag{44.18}$$

Deduce that the smallest eigenvalue (corresponding to a TE mode) is π, divided by the larger of a_x and a_y.

Problem 44–4. Calculate the surface charge and current densities, and the stress, which the modes (44.17) yield on the walls of a rectangular waveguide.

It is a straightforward matter to calculate the z component of the time averaged energy flux $\Re(\frac{1}{2}\mathbf{E} \times \mathbf{H}^*)$ for the various modes. Integration over the cross section then yields the time averaged rates of energy flow through the cross section. The results are:

$$TEM: \quad c \int \epsilon_0 |\mathbf{grad}_2 V|^2 \, dA,$$

$$TM: \quad \frac{\omega k_z}{2\kappa^2} \int \epsilon_0 |E_z|^2 \, dA,$$

$$TE: \quad \frac{\omega k_z}{2\kappa^2} \int \mu_0 |H_z|^2 \, dA. \tag{44.19}$$

Our formulae were obtained on the assumption that the walls were perfectly conducting. Actual metallic walls have finite conductivity and will absorb some of the radiation. To first order in the skin depth of the walls, we may calculate the rate of absorption by substituting the foregoing ideal solutions in (43.13). In this approximation, the walls will absorb energy, per unit length of the waveguide, at the rate $(2\sigma\delta)^{-1} \oint |H_s|^2 \, dl$, where the line integral is taken along the circumference of the cross section.

PROBLEM

Problem 44–5. Show that, per unit length of the wave guide, the rates of absorption are

$$TEM: \quad \frac{1}{2\mu_0\sigma\delta} \oint \epsilon_0 |\mathbf{grad}_2 V|^2 \, dl,$$

$$TM: \quad \frac{\omega^2}{2\mu_0\sigma\delta c^2\kappa^2} \oint \epsilon_0 |\mathbf{grad}_2 E_z|^2 \, dl,$$

$$TE: \quad \frac{1}{2\mu_0\sigma\delta} \oint \mu_0 \left(|H_z|^2 + \frac{k_z^2}{\kappa^2} |\mathbf{grad}_2 H_z|^2 \right) dl. \tag{44.20}$$

Absorption by the walls will cause the energy flow through the cross section of the waveguide to decay as $e^{-\alpha z}$, where – for each mode – the *attenuation coefficient* α is the ratio of the appropriate expression in (44.20) to the corresponding expression in (44.19). The field amplitudes, too, will decay exponentially (with a decay factor per unit length equal to $\alpha/2$).

Waveguides are used for transmitting electromagnetic radiation; for example, from a device that generates the radiation to an antenna. Despite the exponential decay of the energy flow, transmission of radiation along waveguides may – over not too long distances – be far more efficient than the alternative transmission through free space: radiation emanating from a source propagates as a spherical wave, and its intensity decreases as the square of the distance; moreover, the collecting device may be a target that subtends a minute solid angle (compared with 4π).

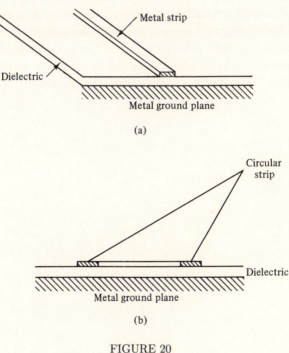

(a)

(b)

FIGURE 20

A waveguide need not always have the form of a pipe. There may be openings in the walls. Consider, for example, a pair of parallel, perfectly conducting planes. These provide a waveguide that allows the propagation of a *TEM* wave in the form of a plane wave, polarized normally to the planes. Practical designs may take the form of two parallel strips, supported by a transparent dielectric slab between them; in this case we must apply the transformation (44.12). An arrangement that allows several such strip lines uses a common ground plane and is shown in Figure 20(a).

If we close the ends of a waveguide, we get a *resonator*. The added boundary conditions at the closed ends will then select a discrete set of wave numbers k_z (and a corresponding discrete set of frequencies ω). It is easily seen that the allowed k_z's are determined by the condition that the distance between the closed ends be an integral number of half-wavelengths. The effect is similar to that of clamping a taut string: instead of travelling along a waveguide, the waves in a resonator are *standing electromagnetic waves*. We may also join the two ends of a waveguide, rather than close each one of them; that would result in a ring-shaped resonator. Again, a resonator need not be closed on all sides. Figure 20(b) shows a circular strip-line resonator.

APPENDIX

Gaussian Units

In the foregoing chapters we have introduced the SI electromagnetic units and used them throughout. It will be recalled that these units were based on the definition of the ampere (cf. (12.7)),

$$\mu_0 = 4\pi\, 10^{-7}\, \frac{\text{newton}}{\text{amp}^2}. \tag{A.1}$$

All other SI electromagnetic units followed from this definition, together with Maxwell's equations and the various defining relations for other quantities, such as $C = Q/V$ or $R = V/i$. In terms of these units, we have also written the relation $\epsilon_0\mu_0 = 1/c^2$ in the form (cf. (12.8))

$$\frac{1}{4\pi\epsilon_0} = \frac{\mu_0 c^2}{4\pi} = 9 \times 10^9\, \frac{\text{newton} \cdot \text{m}^2}{\text{coulomb}^2}, \tag{A.2}$$

where the factor 9 (*not* the exponent 9) stands for the square of the speed of light c in units of 10^8 m/s.

Many other systems of units have been proposed during the past one hundred and fifty years, but only two of them are still widely used: the SI system and the Gaussian system. The SI system is used by all engineers and technicians, and also by many physicists. The trend in the last four decades has certainly been in favor of a universal adoption of the SI system. But the Gaussian system is still being used by physicists, especially in atomic and particle physics. It also appears in some of the most popular textbooks and reference books, and is therefore indispensable to any physicist.

207

We have seen that Maxwell's equations result from the two tensor equations (4.1) and (6.1) after giving the names $\{\mathbf{H}, \mathbf{D}\}$ and $\{\mathbf{E}, \mathbf{B}\}$ to the components of the charge-current potential f and the electromagnetic field F. In the Gaussian system, these names are given differently: the new (primed) names are related to the SI names through the equality of the second and third columns in Table 1. For example, SI charge Q and Gaussian charge Q' are related through $Q = \sqrt{4\pi\epsilon_0}Q'$. Note that the Gaussian quantities have dimensions that are different from those of the corresponding SI quantities. With these new names, Maxwell's equations and the aether relations become

$$\text{div}\,\mathbf{D}' = 4\pi q', \qquad \text{curl}\,\mathbf{H}' = \frac{4\pi}{c}\mathbf{j}' + \frac{1}{c}\mathbf{D}'_t,$$

$$\text{div}\,\mathbf{B}' = 0, \qquad \text{curl}\,\mathbf{E}' = -\frac{1}{c}\mathbf{B}'_t,$$

$$\mathbf{D}' = \mathbf{E}' + 4\pi\mathbf{P}', \qquad \mathbf{H}' = \mathbf{B}' - 4\pi\mathbf{M}'. \qquad (A.3)$$

TABLE 1

Quantity	SI	Gaussian
Charge	$Q\,(q, \sigma)$	$(4\pi\epsilon_0)^{\frac{1}{2}}Q'\,(q', \sigma')$
Current	$i\,(\mathbf{j}, \mathbf{K})$	$(4\pi\epsilon_0)^{\frac{1}{2}}i'\,(\mathbf{j}', \mathbf{K}')$
Charge potential	$\mathbf{D}\,(\mathbf{P})$	$(\epsilon_0/4\pi)^{\frac{1}{2}}\mathbf{D}'\,(4\pi\mathbf{P}')$
Current potential	$\mathbf{H}\,(\mathbf{M})$	$(4\pi\mu_0)^{-\frac{1}{2}}\mathbf{H}'\,(4\pi\mathbf{M}')$
Electric field	$\mathbf{E}\,(V)$	$(4\pi\epsilon_0)^{-\frac{1}{2}}\mathbf{E}'\,(V')$
Magnetic field	$\mathbf{B}\,(\mathbf{A})$	$(\mu_0/4\pi)^{\frac{1}{2}}\mathbf{B}'\,(\mathbf{A}')$

Of course, in the Gaussian system these equations are written without any primes. The most striking fact about these equations is that the Gaussian fields $\mathbf{D}'$, $\mathbf{H}'$, $\mathbf{E}'$, $\mathbf{B}'$, $\mathbf{P}'$ and $\mathbf{M}'$ all have the same dimensions.

We also write down a few other obviously useful relations:

$$\mathcal{E} = \mathbf{E} + \dot{\mathbf{x}} \times \mathbf{B} = \frac{1}{\sqrt{4\pi\epsilon_0}}(\mathbf{E}' + \frac{1}{c}\dot{\mathbf{x}} \times \mathbf{B}') = \frac{1}{\sqrt{4\pi\epsilon_0}}\mathcal{E}',$$

$$\mathcal{H} = \mathbf{H} - \dot{\mathbf{x}} \times \mathbf{D} = \frac{1}{\sqrt{4\pi\mu_0}}(\mathbf{H}' - \frac{1}{c}\dot{\mathbf{x}} \times \mathbf{D}') = \frac{1}{\sqrt{4\pi\mu_0}}\mathcal{H}',$$

$$\mathcal{E} \times \mathcal{H} = \frac{c}{4\pi}\mathcal{E}' \times \mathcal{H}', \qquad \mathcal{J} \cdot \mathcal{E} = \mathcal{J}' \cdot \mathcal{E}',$$

$$\frac{\epsilon_0 E^2}{2} = \frac{E'^2}{8\pi}, \qquad \frac{B^2}{2\mu_0} = \frac{B'^2}{8\pi}. \tag{A.4}$$

Since the electromagnetic quantities in the Gaussian system have different dimensions, their units must be newly defined. There is a difference even in the mechanical units, for the Gaussian system uses CGS mechanical units. It is obvious from (A.2) that the Gaussian charge $Q' = Q/\sqrt{4\pi\epsilon_0}$ has the dimensions of length times the square root of force. The Gaussian unit of electric charge is called the *electrostatic unit* (esu), and is defined as

$$1 \text{ esu} = 1 \text{ dyne}^{1/2} \cdot \text{cm}. \tag{A.5}$$

It follows that

$$\frac{1}{4\pi\epsilon_0} = 9 \times 10^9 \frac{\text{newton} \cdot \text{m}^2}{\text{coulomb}^2}$$

$$= 9 \times 10^{18} \frac{\text{dyne} \cdot \text{cm}^2}{\text{coulomb}^2} = 9 \times 10^{18} \frac{\text{esu}^2}{\text{coulomb}^2}. \tag{A.6}$$

Since the esu and the coulomb have different dimensions, it is meaningless to inquire about 'the number of esu's contained in one coulomb'. What we *can* ask is, if the SI charge Q is 1 coulomb, how many esu's are there in the corresponding Gaussian Q'? In order to answer this question, we set $Q = 1$ coulomb and use the first line of Table 1:

$$Q' = \frac{Q}{\sqrt{4\pi\epsilon_0}} = \frac{3 \times 10^9 \text{ esu}}{\text{coulomb}} 1 \text{ coulomb} = 3 \times 10^9 \text{ esu}. \tag{A.7}$$

This gives the first line in Table 2.

TABLE 2

Quantity	SI	Gaussian
Charge	1 coulomb	3×10^9 esu
Current	1 amp	3×10^9 statamp
Potential	1 volt	$(300)^{-1}$ statvolt
Capacitance	1 farad	9×10^{11} statfarad
Resistance	1 ohm	$(9 \times 10^{11})^{-1}$ statohm
Conductivity	1 siemen/m	9×10^9 /statohm/cm
Electric field	1 volt/m	$(3 \times 10^4)^{-1}$ statvolt/cm
Charge potential	1 coulomb/m^2	$12\pi\, 10^5$ esu/cm^2
Polarization	1 coulomb/m^2	3×10^5 esu/cm^2
Current potential	1 amp/m	$4\pi\, 10^{-3}$ oersted
Magnetic field	1 tesla	10^4 gauss
Magnetic flux	1 weber	10^8 maxwell
Inductance	1 henry	$(9 \times 10^{11})^{-1}$ stathenry

The Gaussian unit of current is 1 statamp=1 esu/s. According to (A.7), an SI current of 1 amp corresponds to a Gaussian current of 3×10^9 statamp.

The Gaussian unit of electric potential, the *statvolt*, is defined by

$$1 \text{ statvolt} = 1\, \frac{\text{esu}}{\text{cm}} = 1\, \frac{\text{dyne} \cdot \text{cm}}{\text{esu}}, \qquad (A.8)$$

the last equality following from (A.5). In order to find the correspondence between statvolts and volts, we set $V = 1$ volt $= 1$ newton· m/coulomb in $V' = V/\sqrt{4\pi\epsilon_0}$ (electric potentials are related like electric fields) and obtain

$$V' = \frac{\text{coulomb}}{3 \times 10^9 \text{ esu}}\, 1\, \frac{\text{newton} \cdot \text{m}}{\text{coulomb}} = \frac{1}{300}\, \frac{\text{dyne} \cdot \text{cm}}{\text{esu}} = \frac{1}{300}\, \text{statvolt.} \quad (A.9)$$

The Gaussian definition of capacitance is

$$C' = \frac{Q'}{V'} = \frac{1}{4\pi\epsilon_0} \frac{Q}{V}. \tag{A.10}$$

Its unit is the *statfarad*, defined by

$$1 \text{ statfarad} = 1 \frac{\text{esu}}{\text{statvolt}} = 1 \text{ cm}. \tag{A.11}$$

Since the statfarad is nothing else than a centimetre, the name merely serves to remind us that we are concerned with a capacitance. If we set $C = 1$ farad $= 1$ coulomb/volt in (A.10), we obtain

$$C' = 9 \times 10^{18} \frac{\text{esu}^2}{\text{coulomb}^2} \frac{\text{coulomb}^2}{\text{newton} \cdot \text{m}}$$

$$= 9 \times 10^{11} \frac{\text{esu}^2}{\text{dyne} \cdot \text{cm}} = 9 \times 10^{11} \text{ statfarad}. \tag{A.12}$$

Gaussian resistance and conductivity are defined by $R' = V'/i'$ and $\mathcal{J}' = \sigma'\mathcal{E}'$. Their units are, respectively,

$$1 \text{ statohm} = 1 \frac{\text{statvolt}}{\text{statamp}} = 1 \frac{\text{s}}{\text{cm}},$$

$$1 (\text{statohm} \cdot \text{cm})^{-1} = 1 \frac{\text{statamp}}{\text{statvolt} \cdot \text{cm}} = 1 \text{ s}^{-1}. \tag{A.13}$$

The correspondences between statohms and ohms, and between (stat-ohm·cm)$^{-1}$ and siemen/m, are easily worked out and are given in the fifth and sixth lines of Table 2.

Next, we consider the fields. The Gaussian electric field $\mathbf{E}'$ has the dimensions of Gaussian electric potential, divided by length. Its unit may be taken as 1 statvolt/cm. Since $\mathbf{E}' = \sqrt{4\pi\epsilon_0}\mathbf{E}$, we find that an SI electric field of 1 volt/m corresponds to

$$E' = \frac{\text{coulomb}}{3 \times 10^9 \text{ esu}} 1 \frac{\text{newton}}{\text{coulomb}} = \frac{1}{3 \times 10^4} \frac{\text{statvolt}}{\text{cm}}; \tag{A.14}$$

the difference between the 3×10^4 in this equation and the 300 in (A.9) results from 1 m $= 100$ cm.

The Gaussian charge potential (or electric displacement) $\mathbf{D}'$ has the dimensions of Gaussian charge, divided by area (cf. (A.3)$_1$). Its unit may be taken as 1 esu/cm^2, which is the same as the electric field unit 1 statvolt/cm (we have already noted that the Gaussian fields all have the same dimensions). Since $\mathbf{D}' = \sqrt{4\pi/\epsilon_0}\mathbf{D}$, an SI charge potential of 1 coulomb/m^2 corresponds to

$$D' = 4\pi\, 3 \times 10^9 \, \frac{\text{esu}}{\text{coulomb}} \, 1 \, \frac{\text{coulomb}}{10^4 \, \text{cm}^2} = 12\pi\, 10^5 \, \frac{\text{esu}}{\text{cm}^2}. \qquad (A.15)$$

The polarization includes an extra factor of 4π (cf. Table 1, or (A.3)$_5$). An SI polarization of 1 coulomb/m^2 therefore corresponds to a Gaussian polarization of 3×10^5 esu/cm.

In the case of the Gaussian current potentials $\mathbf{H}'$ and $\mathbf{M}'$, the field unit of esu/cm^2, or statvolt/cm, is called the *oersted*. According to (A.1) or (A.2) we have

$$\sqrt{4\pi\mu_0} = 4\pi\, 10^{-1} \, \frac{\text{esu}}{\text{amp} \cdot \text{cm}}. \qquad (A.16)$$

Since $\mathbf{H}' = \sqrt{4\pi\mu_0}\mathbf{H}$, an SI charge potential of 1 amp/m corresponds to

$$H' = 4\pi\, 10^{-1} \, \frac{\text{esu}}{\text{amp} \cdot \text{cm}} \, 1 \, \frac{\text{amp}}{10^2 \, \text{cm}} = 4\pi\, 10^{-3} \text{ oersted.} \qquad (A.17)$$

The magnetization (like the polarization) includes an extra factor of 4π. An SI magnetization of 1 amp/m therefore corresponds to a Gaussian magnetization of 10^{-3} oersted.

In the case of the Gaussian magnetic field $\mathbf{B}'$, the field unit of statvolt/cm is called the *gauss*. Since $\mathbf{B}' = \sqrt{4\pi/\mu_0}\mathbf{B}$, a magnetic field of 1 tesla = 1 volt·s/m^2 corresponds to

$$B' = 10 \, \frac{\text{amp} \cdot \text{cm}}{\text{esu}} \, 1 \, \frac{\text{newton} \cdot \text{m} \cdot \text{s}}{\text{coulomb} \cdot \text{m}^2} = 10^4 \text{ gauss.} \qquad (A.18)$$

The Gaussian magnetic flux unit of gauss·cm^2 is called the *maxwell*. Since 1 m^2 = 10^4 cm^2, an SI flux of 1 weber corresponds to a Gaussian flux of 10^8 maxwell.

Finally, we consider the inductance. In the SI system it is defined as magnetic flux, divided by current: $L = \Phi/i$; its unit is 1 henry = 1 weber/amp. The Gaussian inductance is defined by $L' = \Phi'/(ci')$; note the velocity of light in the denominator. Its unit, the *stathenry*, is

$$1 \text{ stathenry} = 1 \frac{\text{maxwell}}{\text{cm} \cdot \text{s}^{-1} \cdot \text{statamp}} = 1 \frac{\text{s}^2}{\text{cm}}. \qquad (A.19)$$

It is easy to show that $L' = 4\pi\epsilon_0 L$. An SI inductance of 1 henry = 1 weber/amp therefore corresponds to a Gaussian inductance of

$$L' = \frac{\text{coulomb}^2}{9 \times 10^{18} \text{ esu}^2} 1 \frac{\text{volt} \cdot \text{s}}{\text{amp}}$$

$$= \frac{\text{coulomb}^2}{9 \times 10^{18} \text{ esu}^2} \frac{\text{newton} \cdot \text{m} \cdot \text{s}^2}{\text{coulomb}^2} = \frac{1}{9 \times 10^{11}} \text{ stathenry.} \qquad (A.20)$$

The Gaussian system is neither superior nor inferior to the SI system. Atomic physicists still show a preference for the Gaussian system, but the reason for this is not obvious, for they seem to think of electric charge in terms of a multiple of the electron's charge, rather than so many esu's. Perhaps they merely prefer to have Maxwell's equations in the form (A.3). After all, many of the greatest works in electromagnetism have used the Gaussian system.† Gaussian adherents have sometimes claimed that the basic Maxwell-Lorenz aether relations $\mathbf{D}' = \mathbf{E}'$ and $\mathbf{H}' = \mathbf{B}'$ for the total (free and bound) fields are the simplest expression for a fundamental property of space-time, certainly simpler than the SI relations $\mathbf{D} = \epsilon_0 \mathbf{E}$ and $\mathbf{H} = \mathbf{B}/\mu_0$ with the dimensional constants ϵ_0 and μ_0. Should we not then recognize these basic identities by assigning the same dimensions (and the same units) to the electromagnetic fields? Here at last we seem to have an *objective* argument in favor of the Gaussian system. Unfortunately, the claim itself is untrue, for we have seen that the aether relations characterize a special class of frames, rather than a property of space-time.

---

† Maxwell himself used the *electromagnetic* system, a variant of the Gaussian.

Index

Material (*continued*)
  magnetic 126–9
  response 12, 31
  time derivative 42
  viscous 71–2
Matrix scalar product 54
Maxwell 213
Maxwell (unit of magnetic flux) 212
Maxwell's equations 11, 15, 19, 21,
  22, 28, 208
  in media 34–8, 208
Maxwell-Lorenz aether relations 22
Meissner's effect 135
Metallic surface 196–9
Milne-Thomson 99
Mirror, perfect 190
Moment
  dipole 33, 91
  magnetic 33
  monopole 91
  multipole 90–3
  quadrupole 92
Momentum 43
  angular 43–4
  generalized 49, 75–6
  kinetic 77
Monochromatic
  conductivity 155
  field 155, 187
  plane wave 185–9
Monopole moment 91
Motor 165
Multipole moments 90–3

Navier-Stokes relations 72
Nernst effect 144
Neumann's problem 89
Noll 2

Oersted (unit of current potential)
  212
Ohm (unit of resistance) 145

Ohm's law 70, 144–5
Ohmic loss 147
Optical fibres 195

Paraelectric 114
Paramagnetic 114, 127
Peltier coefficient 144
Permeability 127
  complex 187
Permittivity 101
  complex 186
Phase velocity 188
Piecewise smooth 14, 21
Piezoelectricity 109–12
Piezomagnetism 128
Plane of incidence 191
Plane of polarization 192
Poisson brackets 76
Poisson's equation 81
Polarization 12, 31, 188
  curve 115
  permanent 113
  plane of 192
  saturation 115
  spontaneous 115, 117
  total 194
  angle of total 194
Ponderable bodies 64
Potential
  advanced 173
  charge-current 9, 11, 23, 31
  complex 97
  dipole 91
  electromagnetic 19
  Gibbs thermodynamic 138
  Lienard-Wiechert 181
  partial 34–8
  regular 86
  retarded 171–3
  scalar 20
  scalar magnetic 122
  thermodynamic 60